René Bonk

MW00814599

**Linear and Nonlinear Semiconductor Optical Amplifiers
for Next-Generation Optical Networks**

Karlsruhe Series in Photonics & Communications, Vol. 8
Edited by Profs. J. Leuthold, W. Freude and C. Koos

Karlsruhe Institute of Technology (KIT)
Institute of Photonics and Quantum Electronics (IPQ)
Germany

Linear and Nonlinear Semiconductor Optical Amplifiers for Next-Generation Optical Networks

by
René Bonk

Dissertation, Karlsruher Institut für Technologie (KIT)
Fakultät für Elektrotechnik und Informationstechnik, 2013

Impressum

Karlsruher Institut für Technologie (KIT)
KIT Scientific Publishing
Straße am Forum 2
D-76131 Karlsruhe
www.ksp.kit.edu

KIT – Universität des Landes Baden-Württemberg und
nationales Forschungszentrum in der Helmholtz-Gemeinschaft

KIT Scientific Publishing 2013
Print on Demand

ISSN 1865-1100
ISBN 978-3-86644-956-5

Linear and Nonlinear Semiconductor Optical Amplifiers for Next-Generation Optical Networks

Zur Erlangung des akademischen Grades eines

DOKTOR-INGENIEURS

von der Fakultät für
Elektrotechnik und Informationstechnik
des Karlsruher Instituts für Technologie (KIT)

genehmigte

DISSERTATION

von

Dipl.-Phys. René Bonk

geb. in: Uelzen

Tag der mündlichen Prüfung: 19.04.2012
Hauptreferent: Prof. Dr. sc. nat. Jürg Leuthold
Korreferenten: Prof. Dr.-Ing. Dr. h. c. Wolfgang Freude
Prof. Dr. rer. nat. Uli Lemmer

To my mother Angelika Bonk,
to my grandmother Margarethe Bachmann,
and in memory of my father Manfred-Eckart Bonk

Table of Contents

List of Figures

Abstract (German)

Die vorliegende Arbeit behandelt die Realisierung, Optimierung und die experimentelle Untersuchung aktiver optischer Komponenten, die für den Bau von zukünftigen optischen hochbitratigen Netzwerken essentiell sind. Zuerst werden die grundlegenden Eigenschaften linearer und nichtlinearer optischer Halbleiterverstärker (engl.: semiconductor optical amplifier, SOA) und deren Einsatz in optischen Regional- und Stadtbereichsnetzwerken der Zukunft untersucht. In Experimenten mit schnellen Datensignalen sowie numerischen Modellen wird die Eingangsdynamik linearer optischer Halbleiterverstärker für deren Anwendung in zukünftigen faserbasierten Zugangsnetzen zur Vergrößerung des tolerierbaren Verlustbudgets, z. B. zur Vergrößerung der Reichweite zwischen Vermittlungsstelle und den Kundenhaushalten, systematisch studiert. In theoretischen Abhandlungen werden für diese SOA die wichtigsten physikalischen Parameter abgeleitet, deren Bedeutung im Hinblick auf die Anwendungen diskutiert und die Bauteile darauffolgend optimiert. Des Weiteren wird durch experimentelle und numerische Untersuchungen gezeigt, dass eine Wellenlängenumsetzung mit optischen Quantenpunktverstärkern auch bei sehr hohen Datenraten unter Verwendung eines speziellen Filters möglich ist. Solche Wellenlängenumsetzer können in optischen Vermittlungsknoten zukünftiger Regionalnetzwerke eingesetzt werden. Die Diskussion eines solchen Knotens mit der neuartigen Funktionalität der optischen Bündelung von Datenströmen beschließt die Arbeit.

In den letzten 20 Jahren hat sich das Internet als Informations- und Kommunikationsinfrastruktur eindrucksvoll durchgesetzt. Menschen wollen heute zu jeder Zeit, an jedem Ort, flexibel auf Internetdienste mit hoher Bandbreite zugreifen können. Diese Entwicklung hatte ein enormes Wachstum der benötigten Bandbreite im Kernnetz zur Folge. Ein Ende dieses Wachstums ist nicht in Sicht. So wurde z. B. von der Firma Cisco Inc. ein Wachstum des Kernnetzdatenverkehrs bis zum Jahr 2015 auf 972 Zettabytes pro Jahr prognostiziert, welches einer jährlichen Wachstumsrate des Gesamtdatenaufkommens von 32 % zwischen 2010 und 2015 entspricht. Mit diesem Wachstum und den steigenden Anforderungen der Kunden ist die Notwendigkeit neuer Kommunikationsnetze mit sehr hoher Bandbreite verbunden. Die benötigte Bandbreite kann nur durch den effizienten Ausbau von optischen Fasernetzen erreicht werden. Den Bedürfnissen angepasste optische Netze werden eine größere Anzahl optischer Elemente benötigen, die einerseits die Leistungsverluste erhöhen, dafür andererseits intelligente Funktionalitäten ermöglichen. Optische Halbleiterverstärker sind vielversprechende, kostengünstige und integrierbare Bauelemente, die in zukünftigen Netzen als Verstärker zur Verlustkompensation wie als nichtlineare Elemente zur Signalverarbeitung eingesetzt werden können.

Lineare optische Halbleiterverstärker (*Abschnitt 4*) sollten einem Datensignal einerseits wenig Rauschen hinzufügen und andererseits nur geringe Signalverzerrungen erzeugen. Das bedeutet, dass ein optischer Halbleiterverstärker eine geringe Rauschzahl und eine hohe Linearität (konstanter Gewinn über großem Eingangsleistungsbereich) aufweisen sollte. Die Verstärkerrauschzahl ist vor allem bei geringen optischen Eingangsleistungen von Bedeutung, während die Linearität, ausgedrückt durch die Sättigungseingangsleistung, bei großen Eingangsleistungen wichtig wird. Der Bereich von optischen Eingangsleistungen, in denen ein Datensignal ohne signifikante Verzerrung verstärkt werden kann, wird Eingangsdynamikbereich (engl.: input power dynamic range, IPDR) genannt. Typischerweise wird der Eingangsdynamikbereich über eine Bitfehlerwahrscheinlichkeit (BER) von 10^{-3} bzw. 10^{-9} am Ausgang des Verstärkers definiert. Je nach Netzanforderung variiert das BER-Kriterium. Ein hoher Eingangsdynamikbereich ist vor allem im Rückkanal eines optischen Zugangsnetzwerkes zwischen den Kundenhaushalten und der Vermittlungsstelle von großer Bedeutung, da durch die unterschiedlichen Entfernungen der Kundenhaushalte zum Verstärker stark unterschiedliche Eingangsleistungen am Verstärker anliegen können.

In der vorliegenden Arbeit wurde erstmalig gezeigt, dass ein großer Dynamikbereich (IPDR) von mehr als 35 dB bei einer Bitfehlerwahrscheinlichkeit von mindestens 10^{-3} mit einer geschickten Wahl der Dimensionalität des Elektronensystems der aktiven SOA-Zone sowie der Bauteilparameter erzielt werden kann. Dabei wurden keine zusätzlichen Hilfsmittel wie Filter oder Laser eingesetzt, die eine Kommerzialisierung durch die entstehenden Zusatzkosten erschweren würden. In den Messungen und Modellbildungen wurden Intensitäts- und Phasenmodulation sowie höherwertige Modulationsformate mit Datenraten von bis zu 80 Gbit/s verwendet.

Ein großer Eingangsdynamikbereich kann durch folgende Parameter erreicht werden: Hohe Besetzungsinversion, großer effektiver Wellenleiterquerschnitt (geringer optischer Feldkonzentrationsfaktor), kleiner differentieller Gewinn, geringe effektive Ladungsträgerlebensdauer, geringe Amplituden-Phasen-Kopplung (kleiner alpha-Faktor), Reduzierung der Dimensionalität des Elektronensystems der aktiven SOA-Zone, geschickte Wahl des Verstärkergewinns. Der Gewinn des Verstärkers sollte immer so gering wie möglich, aber so hoch wie für die Anwendung notwendig gewählt werden.

Da gerade die neuartigen optischen Quantenpunkthalbleiterverstärker eine Vielzahl dieser Parameter aufweisen, wurden diese vielversprechenden nano-photonischen Bauelemente, die selbstorganisierte InAs bzw. InP Quantenpunkte als aktives Medium verwenden, in der vorliegenden Arbeit systematisch untersucht und optimiert. Neben den oben genannten Modulationsformaten kamen auch verschiedene Multiplexverfahren zum Einsatz, um eine höhere Gesamtdatenrate zu erreichen. So wurden z. B. ein Wellenlängen- und ein orthogonales Frequenzmultiplexverfahren untersucht. Die Eignung dieser Bauteile in einer Zugangsnetzstruktur wie auch in einer Kaskade von bis zu 10 Verstärkern wurde studiert. Zusätzlich wurde die Verstärkung von Radiosignalen charakterisiert, die einem optischen Träger aufmoduliert worden waren.

Nichtlineare optische Halbleiterverstärker (*Kapitel 5*) werden in der optischen Signalverarbeitung eingesetzt. In der vorliegenden Arbeit wurde die optische Wellenlängenumsetzung von hochbitratigen intensitätsmodulierten Datensignalen mit einem optischen Halbleiterverstärker untersucht. Das Ziel der Wellenlängenumsetzung ist die Auflösung von Datensignalblockaden beim Herstellen optischer Verbindungen, eine der großen Herausforderungen zukünftiger optischer Kommunikationsnetze. In herkömmlichen optischen Halbleiterverstärkern wird eine Wellenlängenumsetzung eines Datensignals vor allem durch den Effekt der Kreuzphasenmodulation erreicht. Bei Verwendung von hochbitratigen Datensignalen treten allerdings Bitmustereffekte auf, die die Signalqualität nach dem Verstärker limitieren. Typischerweise wird dem optischen Halbleiterverstärkern ein Filter nachgeschaltet, welches zur Reduzierung von Bitmustereffekten führt.

Neuartige optische Quantenpunkthalbleiterverstärker ermöglichen auch bei höheren Datenraten eine Wellenlängenumsetzung ohne Bitmustereffekte. Dieses Verhalten liegt an der schnellen Erholung des Gewinns in diesen Bauteilen, was auf das Nachfüllen der Quantenpunktzustände aus einem Trägerreservoir zurückzuführen ist. Die Wellenlängenumsetzung mit einem optischen Quantenpunkthalbleiterverstärker basiert vor allem auf dem Effekt der Kreuzgewinnmodulation. Bisher nicht untersuchte Fragen welche Filter zur Wellenlängenumsetzung vorteilhaft einsetzbar sind und welche Datenraten erreicht werden können, konnten in dieser Arbeit beantwortet werden.

In Experimenten und Simulationen mit intensitätsmodulierten Signalen von bis zu 80 Gbit/s wurden optimale Filter für die Wellenlängenumsetzung mit Quantenpunktverstärkern ermittelt. Eine hohe Signalqualität der Wellenlängenumsetzung mit optischen Quantenpunktverstärkern und nachgeschaltetem Filter konnte durch Messung einer Bitfehlerwahrscheinlichkeit von 10^{-9} bei einer Datenrate von 80 Gbit/s nachgewiesen werden.

Schließlich berichtet *Kapitel 6* über einen *neuartigen optischen Vermittlungsknoten* zur speicherfreien Verbindung eines Stadtringnetzes und eines Zugangsringnetzes. Dieser Knoten weist Schaltfunktionalitäten im Zeit-, Raum-, und Frequenzbereich auf, welche experimentell untersucht und demonstriert wurden. Neuartige Konzepte wurden für die Kernfunktionalitäten des Knotens entwickelt: Bündelung von Datenströmen mit Zeitschlitzvertauschung, Umsetzung von Wellenlängenmultiplex in optischen Zeitmultiplex und umgekehrt, sowie die gleichzeitige Regeneration mehrerer hochbitratiger Datensignale.

Preface

Novel nano-photonic devices are key elements to satisfy the demand for bandwidth and numerous services. Next-generation high-speed all-optical networks will use semiconductor optical amplifiers (SOA) to overcome losses induced by the huge amount of intelligent network components and the long reach of the fiber-based networks. Additionally, all-optical signal processing devices are required for photonic switching, wavelength conversion or clock recovery. Thus, compact, fast, linear and nonlinear devices such as SOA are the technology of choice for future high-speed optical networks.

By the year 2015, the global IP traffic is predicted to be equivalent to about 250 billion DVDs per year. The overall IP traffic is growing with a compound annual growth rate of around 30 % in the next years (Cisco Systems Inc.). The growth results from new applications and services such as internet video, IP video-on-demand, or video conferencing. It is further very interesting to note that the internet traffic generated from wireless devices will strongly increase. Obviously, to satisfy the demand of customers in the near and also in the far future, flexible, reconfigurable, long-reach, non-blocking, high-speed photonic networks are required.

While much of the recent years' research and development has been focussed on the implementation of high-capacity backbone networks, the speed of metro networks and access networks are still lagging behind. Legacy metro-ring networks use electrical cross-connects and electrical switches. Thus, current metro networks carry data inefficiently, resulting in a bandwidth bottleneck. This bandwidth bottleneck, which is also known as the metro gap, prevents high-speed clients and service providers in local access networks from using the vast amount of bandwidth available in the high-capacity wavelength-division-multiplexing (WDM) backbone networks.

At present, there is a strong worldwide push towards bringing the fiber closer to individual homes and businesses with the goal to overcome the last mile issue. Future access networks have to support a wide range of new services and applications with bit rates of up to 1...10 Gbit/s per customer. Thus, intelligent all-optical network elements will be deployed in the access and in the metro network to provide this bandwidth. However, an increasing amount of network elements will also increase the total network loss. Additionally, there is a need for intelligent signal processing devices to switch and wavelength convert data signals in the optical domain.

SOA are good candidates for enabling high-speed optical amplification and also high-speed optical signal processing. Especially, novel quantum-dot (QD) SOA can offer a large gain bandwidth exceeding 100 nm, they can be operated without bit-pattern effects and they can show low amplitude-to-phase coupling if operated properly.

In this work, applications of optical components for next-generation optical networks are investigated. First, SOA are studied in the linear gain regime in which the gain is constant.

Parameters are derived to increase this linear gain regime. Since the optimum specifications favor QD SOA, they are used for the amplification of signals with different modulation formats and multiplexing techniques. Second, SOA are studied in their nonlinear gain regime in which the gain is suppressed. Parameters are derived to optimize the SOA performance in this nonlinear gain regime. The derived optimum specifications favor conventional SOA. However, since QD SOA can show a very fast gain recovery it is interesting to investigate their suitability for nonlinear signal processing of, e.g., intensity-encoded data signals. As an example, a wavelength converter is studied. A wavelength converter is an important device to overcome wavelength blocking issues in next-generation optical switches. Third, such an all-optical switch connecting a metro-core ring network with metro-access ring networks is realized and experimentally demonstrated.

This thesis is structured as follows.
In *Chapter 1*, the motivation of using optical components, especially SOA, in next-generation all-optical networks is given. Network requirements, topologies and the key driving applications are discussed.
In *Chapter 2*, the SOA basics are presented. Parameters describing the SOA performance are derived. Additionally, basics of QD SOA are introduced, because some features of these devices cannot be explained with the theory of conventional SOA.
In *Chapter 3*, the steady-state and the dynamic properties for all SOA used in this thesis are presented.
In *Chapter 4*, the input power dynamic range (IPDR) of SOA is investigated. The IPDR defines the power range in which error-free amplification can be achieved. Design guidelines to maximize the IPDR with record values exceeding 32 dB for intensity-modulated 40 Gbit/s signals at a given gain are provided and experimentally verified. Further, the capability of SOA to amplify signals with advanced optical modulation formats is investigated. Especially the impact of the SOA alpha factor as a measure of amplitude-phase coupling is studied for such signals.

Applications of SOA with large IPDR used as reach-extender amplifiers for next-generation access networks, for radio-over-fiber networks and as loss-compensating elements in reconfigurable optical add-drop multiplexers (ROADM) are demonstrated.
In *Chapter 5*, the nonlinear operation regimes of QD SOA are investigated and the ideal filter providing the best all-optical wavelength conversion efficiency is derived theoretically. These results are confirmed by experiments with up to 80 Gbit/s on-off keying (OOK) signals using a QD SOA in combination with a delay-interferometer filter.

Additionally, a single and multi-wavelength clock recovery circuit using a filter and a QD SOA is studied at 40 Gbit/s OOK.
In *Chapter 6*, a novel regenerative all-optical grooming switch for interconnecting a 130 Gbit/s OOK metro-core ring and 43 Gbit/s OOK metro-access rings with switching functionality in time, space and wavelength domain is reported.
In *Chapter 7*, a summary of the work and an outlook for future research are provided.

Achievements of the Present Work

In this thesis, optical components for next-generation optical networks were optimized, realized and experimentally tested. First, semiconductor optical amplifiers (SOA) were investigated in their linear gain regime in which the gain is constant, see Fig. 2.11. Here, SOA were used to amplify single channel data signals with different modulation formats and multi-channel signals multiplexed with different techniques, see *Chapter 4*. Second, SOA were studied in their nonlinear gain regime in which the gain is suppressed, see Fig. 2.11. Here, SOA were used for nonlinear signal processing of intensity-modulated data signals, see *Chapter 5*. Third, a novel all-optical grooming switch connecting a metro-core ring network with metro-access ring networks was realized, see *Chapter 6*.

In the following, a brief summary of the main achievements is provided:

Linear SOA (*Chapter 4*)
Optimization of SOA Design Parameters for Large Input Power Dynamic Range (IPDR):
A description relating SOA device parameters to the IPDR is given (*Section 4.3.1, page 82*). The IPDR is large if the SOA has a low noise figure and a high saturation power, i.e., if it is linear. These specifications favor QD SOA over other SOA types (*Fig. 4.12, Section 4.3.7.2, page 94*, [J2], [B1]).

Linear Amplification and IPDR of Signals having Different Modulation Formats:
The IPDR of SOA has been studied in general. For QD SOA, record-high IPDR values of 32 dB at a gain > 15 dB for a target BER of 10^{-9} are reported for on-off keying (OOK) data signals at 40 Gbit/s (*Fig.* 4.9, *Section 4.3.6, page 91*, [J2]). Applications of such linear SOA are predicted as reach-extender amplifier for next-generation access networks (*Section 4.6*, [C25]) and for radio-over-fiber networks (*Section 4.5*, [C9]).

For the first time, it is shown that an SOA with a low alpha factor offers advantages when advanced optical modulation formats are not too complex (*Fig. 4.20, Section 4.4.2.3 page 107*, [J1], [J6], [C4], [C19]).

Concatenation of Reconfigurable Optical Add-drop Multiplexer (ROADM) Containing SOA for Orthogonal Frequency-division-multiplexing (OFDM) Signals:
Recirculating fiber-loop experiments with one QD SOA are performed for the first time with intensity-modulated and direct-detected OFDM signals with BPSK (4.5 Gbit/s including overhead) and QPSK (9 Gbit/s including overhead) modulated subcarriers. Rules for the range of optical channel powers of this SOA cascade were extracted (*Fig. 4.39, Section 4.6.3, page 125*).

viii

Nonlinear SOA (*Chapter 5*)

Wavelength Conversion of High-Speed Data Signals: Record 80 Gbit/s OOK all-optical wavelength conversion using a QD SOA were demonstrated in 2011 (*Fig. 5.21, Section 5.3.2.3, page 159*, [J7], [C28], [C35]).

Grooming Switch (*Chapter 6*)

Novel All-Optical Grooming Switch: A node with grooming functionality and multi-wavelength regeneration was successfully demonstrated. The node connects a 130 Gbit/s metro-core ring with a number of 43 Gbit/s metro-access rings. It offers switching functionality in time, i.e., including time-slot interchanging, space and wavelength domain (*Fig. 6.4, Section 6.4, page 178*, [J5], [J8], [J12]).

1 Introduction

The evolution of the internet as a global information system has changed the social and economical life over the last 20 years. Video services, online gaming, voice over internet protocol (IP) and high definition television (HDTV) are examples of services which customers are using. These services cause a large amount of internet traffic. It is forecast by Cisco Systems Inc. [1] that the global internet traffic will grow to 972 exabytes per year by the year 2015. This corresponds to a compound annual growth rate of 32 percent, and means that the global IP traffic will quadruple from 2010 to 2015. The IP traffic growth will significantly change the required optical network topology, bandwidth, and service overlays. The realization of versatile next-generation optical networks needs new devices and technologies based on, e.g., indium phosphide (InP), gallium arsenide (GaAs) or silicon (Si) semiconductors. These semiconductors can be used to generate high-speed laser sources, modulators, amplifiers, receivers and switches. In recent years, it has been predicted that the semiconductor optical amplifier (SOA) will be a key device of next-generation optical networks. SOA have attracted much interest for two reasons. First, SOA have the ability to amplify signals across the whole spectral range in 10 THz windows from 1250 nm to 1600 nm at reasonable costs. Second, SOA can be successfully used for nonlinear signal processing.

In this chapter, a brief introduction to the IP traffic forecast from 2010 to 2015 is presented in *Section 1.1*, and current access and metro network implementations are introduced in *Section 1.2*. This part of the chapter can be considered standard knowledge and is a subject of various textbooks. Comprehensive presentations of legacy metro and access networks can be found in [2]-[6]. Then, requirements of future optical networks are presented in *Section 1.3*. Finally, in *Section 1.4*, the main focus of this work is introduced which is the integration of SOA reach extenders, SOA loss compensators, SOA wavelength converters, SOA clock recovery circuits and optical switches into a converged metro and access network scenario.

1.1 Forecast: IP Traffic Generates Zettabytes by 2015

The discussion in this section follows the forecast of Cisco Systems Inc. on "Entering the Zettabyte Era" [7] and the "Cisco Visual Networking Index: Forecast and Methodology, 2010-2015" [1]. In the next few years, the global IP traffic is expected to grow with a compound annual growth rate of 32 %, see Fig. 1.1. Most of the 81 exabytes per month in 2015 results from internet traffic and managed IP traffic. Internet traffic is traffic crossing internet backbones, and managed IP traffic is traffic generated by transport of TV and video-on-demand (VoD) signals. Additionally, the amount of mobile data is significantly increasing with a compound annual growth rate of 92 %. Mobile data is traffic generated by handsets, notebook cards, and mobile broadband gateways. Most of the total IP traffic in 2015 will be generated by IP video which includes internet video, VoD and video conferencing. The traffic from an application such as voice-over-IP (VoIP) will be almost negligible.

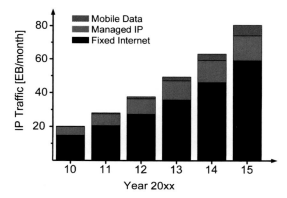

Fig. 1.1: In the next few years, the global IP traffic is expected to grow with a compound annual growth rate of 32 %. The global IP traffic grows from 20 exabytes per month in 2010 to 81 exabytes per month in 2015. The total IP traffic comes from fixed internet data, managed IP data and mobile data.

Residential customers will induce a large downstream component due to the increasing amount of IP traffic from video services. The overall service demands will make the network traffic more asymmetric since the upstream component is not large, except for video calling. It is expected that the average global residential internet connection download speed will grow fourfold to 28 Mbit/s by 2015.

On the contrary, for business customers a more symmetrical network will be required since business IP traffic will grow due to applications such as video communications.

1.2 Review of Existing Access and Metro Networks

This section is considered standard-knowledge, and thus the discussion follows very closely the books and papers [2]-[6], [8]. A common state-of-the-art optical transport network with its different coverage areas is shown in Fig. 1.2. The network planes represent different levels of aggregated traffic and inter-node distances.

A wide-area network (WAN, global backbone, long-haul network) is a network, that spans over large geographical areas and that carries aggregated global network traffic. In optical communications it represents any terrestrial or undersea system. The reach in such networks is between hundreds to thousands of kilometers. The topology in WAN is typically a mesh. In a mesh topology, every node has a circuit connecting it to neighboring nodes. Mesh topologies are very expensive to implement but yield the greatest amount of redundancy. The bit rate between two WAN nodes is about a few Tbit/s. One standard single-mode fiber (SMF: transport medium in WAN) carries several wavelength channels (e.g., up to 80 wavelength channels) at 10 Gbit/s, 40 Gbit/s or even 100 Gbit/s. The channels are spaced (separation of centre carrier frequencies) by 25 GHz...250 GHz, and form a so-called dense wavelength-division-multiplexing (DWDM), see Fig.A. 4 in the *Appendix A.4*.

It should be mentioned that in this work the term optical network is called all-optical when optically transparent networks are meant, and the optical signals are transmitted without any intermediate optical-to-electrical-to-optical (OEO) conversion. Otherwise, the optical network is opaque and referred to without a special attribute.

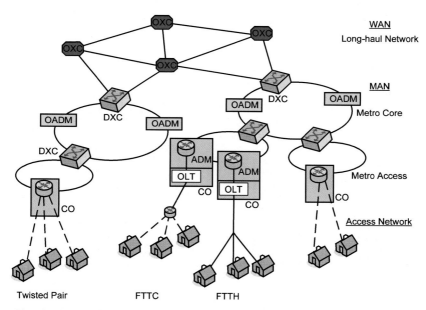

Fig. 1.2: Schematic of common state-of-the-art optical transport network with its different coverage areas from long-haul to access networks. A wide area network (WAN) has typically a mesh topology, and it connects terrestrial or undersea systems. The reach of such networks is between hundreds to thousands of kilometers. The metropolitan area networks (MAN, metro) comprise metro-core rings and metro-access rings. The typical circumference of a metro-core ring is 100-200 km, and the typical circumference of a metro-access ring is 30-60 km. The last-mile network, close to the user premises, is called access network. In WAN several wavelength channels are transmitted in a single fiber, and switching of the data signals is done by means of optical cross-connects (OXC). In the metro networks mostly slower and energy-inefficient digital cross-connects (DXC) are used to aggregate and switch the traffic from the WAN to the MAN and also to switch data between the metro-core and the metro-access rings. Optical add-drop multiplexer (OADM) and electrical add-drop multiplexer (ADM) allow a number of channels to bypass a node while only a few channels will be dropped. In the access network, the data signals are sent from the central offices (CO) to the customers and vice versa. Here, fiber and also copper cables are used to transmit the signals, and the maximum physical reach is 20 km.

Core switches, which perform switching and routing functions, are typically implemented using optical cross-connects (OXC). OXC interconnect an arbitrary number of input ports with an arbitrary number of output ports. N_f fibers are input to an OXC and then the signal in each fiber is demultiplexed onto M_f separate wavelength channels. Afterwards the M_f channels are guided into an electrical or optical switch fabric and directed onto the desired output port. The switch fabric in the OXC is most likely optical. If it is an electrical switch fabric, it is understood that no further electrical multiplexing or signal processing is done. The fabric only performs switching of an input signal into an output port, the same way as an optical switch fabric would do.

In WDM networks an OXC can introduce blocking issues due to wavelength contention, see Fig. 1.3(a). For instance, due to the nature of WDM, a node can only switch wavelength channels to one desired output port (same fiber). This blocking issue can be overcome in wavelength-interchanging OXC in which any wavelength channel can be switched to any output port, see Fig. 1.3(b). This needs a wavelength converter either operating in the optical or in the electrical domain.

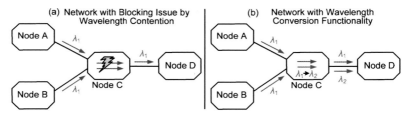

Fig. 1.3: Network scenario without and with wavelength conversion functionality. (a) A node can only multiplex different wavelengths onto a desired output port. If identical wavelength channels have to be multiplexed onto the same output port, blocking occurs due to wavelength contention. This issue can be overcome by using wavelength converters which can be integrated within the node (b).

A metropolitan-area network (MAN, metro) is topologically situated between the WAN and the access network plane. It serves the service providers to transport traffic within their own network and serves as an access network for the WAN. The most common metro network topology is a bidirectional ring due to the inherent failure protection. MAN or metro networks are subdivided into:

Metro core: A ring network with a typical circumference of 100-200 km which carries data on several wavelength channels (e.g., 40 wavelength channels) with a bit rate of 2.5 Gbit/s or 10 Gbit/s in a DWDM spacing. The transport medium is the SMF. Edge routers are used to perform traffic aggregation, switching and routing between the WAN and the metro-core ring. Aggregation and switching are also known as grooming. The edge routers can be realized using digital cross-connect switches (DXC). In the DXC, optical signals are converted into the electronic domain, electrically switched, and then converted back into an optical signal. Typical DXC has N_f input fibers. Each of the N_f input fibers carries M_f optical wavelength channels, which are first demultiplexed onto M_f separate channels. Since electronic switch fabrics process data at 2.5 Gbit/s [3] and up to 40 Gbit/s [9], further electrical demultiplexing could be required after the optical receiver. The advantage of grooming is that channels can be redistributed with a fine granularity (channel capacity / baseband data rate). The disadvantage is that, e.g., a single 40 Gbit/s channel can require the handling and electronic switching of up to 16 individual lower bit rate channels. DXC offers re-timing, re-shaping and re-amplification of the data signal. This is known as regeneration. Add-drop multiplexers (ADM) allow a number of channels to bypass a node while only a few channels will be dropped. Of particular importance are reconfigurable optical add-drop multiplexers (ROADM), where optical signals that bypass the switch stay in the optical domain and a certain number of channels can be dropped or added.

Metro access: A ring network with a typical circumference of 30-60 km which carries a few wavelength channels (e.g., 1...16 wavelength channels) with a bit rate of 0.622 Gbit/s...1 Gbit/s in DWDM or coarse WDM (CWDM, 20 nm carrier wavelength spacing). The transport medium is the SMF. DXC are used to perform traffic aggregation and switching between the metro-core ring and the metro-access ring. At the edge to the access network, an electronic ADM aggregates low bit rate traffic coming from the, e.g., local area networks (LAN), enterprise clients or residential customers.

The access network is the part of the public switched network that connects a central office (CO) or the point of presence with the customer premise equipment (CPE) of subscribers. For instance, the access network includes the subscriber loops and the fiber-to-the-x (FTTx) access networks (x stands for C: Curb, B: Building or H: Home). The traditional transport medium in access networks is the twisted pair (TP) copper cable, known as the subscriber loop. The subscriber loop might be an analog subscriber line or a digital subscriber line (DSL). In recent years, the fiber penetrates towards the CPE with the FFTx approach. Here, in an optical line terminal (OLT), the electrical signals are converted into the optical domain and then send to the CPE. Current maximum downstream bit rates at the customer's side are up to 100 Mbit/s for copper cables and up to 1 Gbit/s for the FTTH approach.

1.2.1 Access Network Implementations

This subsection follows closely the discussion in the books and papers [4], [5], [8]. Today, advanced DSL techniques are deployed using twisted pair and coaxial cable modem techniques. These access networks based on copper as the transmission medium are currently reaching their capacity limits. The higher capacity in copper-based networks can only be achieved at the expense of shorter reach or an increased complexity. Thus, a significant increase in reach and capacity is only achievable with optical fibers. Fiber-based networks offer an inherently huge bandwidth, and low losses. Thus, the trend is to replace the old copper cables by optical fibers in the feeder line from the CO to the customer premises. Currently most of the access networks employ both fiber and copper cables. Fibers connect the CO with a cabinet put at the street curb. Here, the fiber is handed over to the copper-cable network. These so-called FTTC solutions will be replaced in the near future by purely fiber-based optical access networks. This approach is the so-called FTTH.

ber-Cable Access Networks (FTTC, Fiber DSL Networks):

In the FTTC approach, a feeder fiber is used to connect the CO, i.e., the OLT, with a cabinet housing at the curb site, see Fig. 1.4(a). There, a number of modems are placed, where each modem supports a point-to-point connection to a customer premise. Twisted copper pairs can still enable these connections from the modems to the user locations for asymmetric digital subscriber line (ADSL) techniques with a typical length of about a few kilometers (typical bit rate per user about 6 Mbit/s). Additionally, FTTC can also be used for very high-capacity digital subscriber line (VDSL) techniques with a typically reach of some hundreds of meters (typical bit rate per user about 50 Mbit/s). These copper lines suffer from bandwidth limitations and from the cross-talk between the copper pairs as multiple lines are combined in a single cable [8]. These effects are counteracted by the latest innovations in the VDSL technology utilizing bonding and vectoring options (offering up to 100 Mbit/s per subscriber).

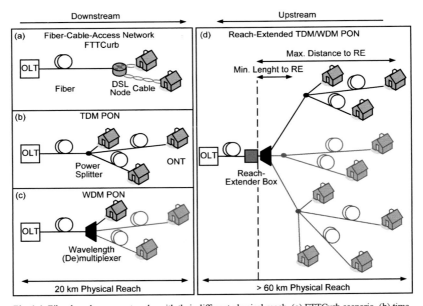

Fig. 1.4: Fiber-based access networks with their different physical reach. (a) FTTCurb scenario, (b) time-division multiplexing (TDM) passive optical network (PON), (c) wavelength-division-multiplexing (WDM) PON and (d) reach-extended hybrid WDM/TDM PON.

Fiber-to-the-Home (FTTH):

High capacity connections exceeding 100 Mbit/s per home can be realized with a network based on fibers all the way to the home. Several topologies can be implemented which are discussed in the following.

Point-to-Point FTTH:

In the point-to-point (P2P) FTTH approach an individual fiber runs all the way from the CO to the user locations. This way there is no competition for capacity among different users. If a single user needs a bandwidth increase, only one specific user needs an upgrade. Moreover, simple Ethernet P2P transceivers can be used, which are becoming very cheap. However, there is a lot of such line-terminating equipment needed, not only at the homes, but also in the CO, which requires more effort for housing, powering and cooling. Moreover, also many fibers need to be installed in the field.

Point-to-Multipoint FTTH:

Savings on the feeder part of the fiber infrastructure can be obtained by deploying a point-to-multipoint (P2MP) topology, thus sharing among the users the major part of the infrastructure, namely the feeder line. A fully passive point-to-multipoint topology is the so-called passive optical network (PON) shown in Fig. 1.4(b). It offers the advantage of sharing the feeder fiber and the line-terminating equipment in the CO. Thus important cost savings can be made on the installation and maintenance. As there is no active equipment in the field, there is no need for powering, for equipment which has to be able to withstand large temperature variations, and for expensive street cabinets. Only an optical power splitter is needed in order to distribute the light signal from the CO to the homes. The 1:x power splitters equally distribute the power from one input port to x output ports. Currently, power splitters with a split ratio of 1:16 (split losses of 12 dB + 1 dB insertion loss per port) are in use. In the future, 1:512 (split losses of 27 dB + 1 dB insertion loss per port) or even higher split ratios are expected.

In PONs two access techniques are currently used for sending and receiving data signals.

Time-division-multiplexing (TDM) PON:

The first technique is the TDM PON. In a TDM PON the capacity of the feeder fiber can be shared by multiple homes. In the downstream direction, i.e., from the CO to the customer homes, each home receives all the signals sent by the CO, and, with an appropriate mechanism, has to pick from these only the signal which is addressed to that particular home. In the upstream direction, the process is a bit more complicated. As the clients may want to send information at random times, a mechanism is needed which avoids collision of all these information streams when they arrive at the power combiner before they enter the shared feeder fiber. Hence in the access network there is a need to deploy an appropriate medium access control (MAC) protocol to give each home a fair share of the feeder's capacity.

In the downstream direction, typically operated at a wavelength around 1490 nm, the OLT at the CO sends the data packets over the PON feeder fiber and thus broadcasts these to each home, see Fig. 1.4(b). Upon receipt by the optical network termination (ONT) at each home, the address field of the packets is inspected. If matching with the address of the home, the packet is accepted and delivered to the user.

For communication in the upstream direction, typically operated at a wavelength around 1310 nm, a time-slot is assigned to each home in which it is allowed to send its packets. By carefully synchronizing these timeslots, collision of the packets at the optical power combiner is avoided. In order to assign the right amount of capacity to each home, the ONT may send first a request packet to the local exchange indicating how much capacity it needs, on which the local exchange may grant one (or multiple) time-slots to the ONT. Thus the upstream capacity per ONT can be adapted to its actual traffic load.

The synchronization needed among the upstream time-slots requires that the propagation time from each ONT up to the OLT is precisely known. These times are generally different, as each ONT is at a different distance from the OLT. They can be determined by sending ranging grants from the OLT, and measuring the roundtrip time upon receipt of these ranging grants returned by each ONT. By inserting appropriate hold-off times at each ONT in order to equalize the length differences, each ONT is put at virtually the same length from the OLT, and can then easily insert its packets in the right time-slot. In addition, the upstream packets from different ONTs arrive at the OLT with different intensity levels depending on the different fiber link losses. The receiver in the OLT therefore needs to operate on packets arriving in burst mode, with different intensity levels, requiring a fast clock extraction and decision level setting per packet. Even though, the processes to be implemented in the OLT and the ONTs for time-division-multiplexing access (TDMA) are not simple, they can be implemented mostly in digital electronics and thus realized at (relatively) low costs.

Hence, TDMA is the most popular MAC protocol for PONs up to now and has been implemented in the standardized schemes for broadband PON (BPON), Ethernet PON (EPON), and gigabit PON (GPON). In GPON downstream speeds of 2.5 Gbit/s (about 100 Mbit/s per subscriber) with on-off keying (OOK) modulation are typically used [8]. The maximum reach of such TDMA GPONs are 20 km with a total acceptable loss budget of 28 dB (Class B+) [10]. In recent years, next-generation PON1 (NG-PON1) access systems are developed which are based on a TDM-PON approach with a downstream bit rate of 10 Gbit/s. Such NG-PON1 networks can offer up to 1 Gbit/s per subscriber in the downstream direction.

Wavelength-division-multiplexing (WDM) PON:
The second solution for realizing independent upstream channels is by using wavelength-division-multiplexing. Each subscriber uses a specific wavelength channel, and in the splitter point of the PON, a wavelength multiplexer/demultiplexer is used to combine/separate these wavelength channels into/from the feeder fiber, see Fig. 1.4(c). Due to the independance of the data channels no synchronization between them is needed, such that they can transport signals with very different formats. As each wavelength channel constitutes an independent path from the CO to a customer location, WDM PON actually offers point-to-point connection functionalities on a shared point-to-multipoint physical infrastructure. Moreover, as the WDM multiplexer performs wavelength routing, the losses in the network splitter point are much lower than in the power splitting PON. For example, in a 1:32 power splitter the losses are about 15 dB, whereas in a 32 channel WDM device the insertion loss can be less than 3 dB per port. Hence the link power budget is considerably better, and thus the reach of a

WDM PON can be larger. As each channel is operating at a specific wavelength, the line-terminating equipment in the OLT at the CO and in the ONT at the user homes needs a wavelength-specific transmitter. This implies that the network operator needs to keep an expensive stock of wavelength-specific modules. A more convenient solution is to deploy universal colorless transceiver modules. Such modules can be realized in various ways. For example, a reflective modulator or reflective semiconductor optical amplifier (RSOA) can be used in the transmitter module at the ONT to generate the upstream data. Upstream data speeds of 1.25 Gbit/s and beyond have been obtained [4], [5], [8].

Other access schemes are also investigated in research and development, e.g., the subcarrier multiple access (SCMA) [5], the orthogonal frequency division multiple access (OFDMA) [11], and the optical code division multiple access (OCDMA) [4].

Next-Generation Reach-Extended PON:
A current trend in research and development is the convergence of TDM and WDM access schemes into a so-called hybrid WDM/TDM PON (e.g., NG-PON2 approach using 4 WDM channels with 10 Gbit/s each in the downstream direction). Thereby, a large-scale FTTH network can be obtained. By deploying multiple wavelength channels a large capacity is realized, and by using TDM within each wavelength channel a large amount of users can be achieved. Such hybrid WDM/TDM schemes are being investigated for long-reach PON systems featuring a reach of more than 100 km, and more than 1000 homes connected, see Fig. 1.4(d). Thus, the number of COs can be reduced, saving operational cost. The longer reach with a higher split ratio causes a loss increase. Thus, reach-extender (RE) amplifiers are required in the downstream and in the upstream paths. Especially in the upstream path, the RE box needs to cope with fluctuating input power levels due to the different distances to the customer locations. The maximum input power variation is determined by the power difference between the ONT with the maximum distance to the RE and the ONT with the minimum distance to the RE. In this thesis, it is assumed that identical split ratios, identical laser diodes in the ONTs (with up to 5 dBm output power in Class B + GPON [12]), and no additionally power leveling protocols are used. The input power variation in today's TDM PON is about 10...15 dB [10]. However, this power variation will increase significantly with an increase in network complexity. It is expected that input power dynamics in the range of 40 dB are needed in next-generation access networks.

In next-generation long-reach PONs, the RE will also have to cope with different high-speed wavelength channels simultaneously, and with advanced optical modulation formats. Overall requirements for a RE box are that it needs to withstand summer heat as well as winter cold, when it is deployed in the street cabinet. Capital expenditures (CAPEX) and operational expenditures (OPEX) should be as low as possible to enable quick revenue.

In the following, a brief overview of possible reach-extender technologies is given, and the respective advantages and disadvantages are discussed.

Semiconductor Optical Amplifier (SOA): An SOA is a semiconductor waveguide with a gain medium. In an SOA the gain is obtained by injecting carriers from a current source into the active region. The SOA technology can offer a large gain bandwidth about 50 nm...100 nm and almost polarization independent amplification. SOA are available in the spectral range of the downstream (around 1.5 µm) and of the upstream (around 1.3 µm), respectively. If the devices are operated in gain saturation, bit-pattern effects can occur for high-speed data signals, and cross-talk in multi-wavelength operation. On the contrary, if the devices are operated with low input power signals, the optical signal-to-noise ratio (OSNR) limits the performance. Here, a large dynamic range is desirable to overcome the limitations. SOA can be deployed in a PON within the CO to boost/pre-amplify the data signal which is sent to/from the customers, as an in-line amplifier in the field or within the ONTs. More details on the basics of SOA are presented in *Chapter 2*.

Raman-Amplifier: If two optical fields are launched into a fiber with a difference frequency that matches the optical phonon resonance frequency, a beating of the fields can lead to strong coherent optical phonon oscillations. The process leads to stimulated scattering of the strong light signal into the weak one. This stimulated Raman scattering can be exploited to amplify an optical signal. In practice, a strong pump laser (located at the CO) at a wavelength of 1450 nm is added to the optical signal which is typically at 1550 nm. As both signals have a spacing of 100 nm or 13 THz in terms of frequencies, energy is transferred. The distributed amplification takes place along a distance of 20 km. The advantages of this technology are that bit-pattern effects can be avoided, that the occurrence of fiber nonlinearities are significantly reduced due to the distributed amplification process, and a low noise figure is possible. However, a low bandwidth in the range of 20 nm (using a single laser source) and the need for a strong pump laser increases the costs compared to the SOA technology. At a wavelength of 1.3 µm this technology is not as mature as the SOA technology.

x-Doped Fiber Amplifier (xDFA): Fiber amplifiers are usually based on rare-earth doped fibers (Erbium (EDFA) for C-band operation or Praseodymium (PDFA) for O-band operation). In contrast to electrically pumped SOA, fiber amplifiers do not only need a doped fiber but also an optical pump laser. Their gain bandwidth is limited to something between 30 nm...70 nm. Especially in the upstream path of access networks, where the traffic is bursty (bursts on millisecond and microsecond time scales), fiber amplifiers induce strong signal degradations due to their fluorescent lifetimes (10 ms in an EDFA). Burst capable EDFAs are available, but they need additional equipment [13]. An advantage of fiber amplifiers is a large small-signal gain of up to 40 dB and low noise figures down to 4 dB or less. They are polarization independent and widely used in multi-wavelength applications due to their low channel cross-talk. Patterning effects at high speed applications (10 Gbit/s and higher bit rates) are avoided due to the long fluorescent lifetimes. *x*DFA are available at moderate cost. At a wavelength of 1.3 µm this technology is no mass product.

Optical-Electrical-Optical (OEO) Units: In active access networks the reach can be extended with OEO units. OEO conversion has already been introduced in the discussion of DXC. The advantages of an OEO conversion are that data signals can be format-transparently regenerated and errors are corrected. However, disadvantages are that an OEO unit is expensive, complex, has a large footprint, and large power consumption. If the bandwidth of an access network requires an upgrade, the OEO units need to be replaced, which is a time consuming and traffic-disruptive procedure.

Coherent Transceiver: The use of phase-modulated signals in conjunction with coherent reception at the OLT and ONT sites can significantly increase the access network reach. The overall power budget is increased due to the fact that lower receiver input power levels are tolerable compared to direct-detection schemes [14].

1.3 Requirements for Next–Generation Optical Networks

In this section, requirements for next-generation all-optical networks in terms of data signal modulation formats, multiplexing techniques, bit rate, and topology are discussed.

The trend towards higher subscriber traffic requires a significant increase of the *bit rate* per wavelength channel and an efficient use of the total fiber bandwidth. There are many ways to increase the bit rate on a single optical carrier wavelength. On one hand, it is possible to increase the speed of intensity-modulated data signals, i.e., OOK data signals. On the other hand, it is possible to use *advanced optical modulation formats*, which are already common in radio-frequency and wireless networks. These higher order optical modulation formats encode the data in amplitude and phase of the optical field [15], see *Appendix A.1*. Examples of higher order modulation formats are phase shift keying (PSK) or M-ary quadrature amplitude modulation (QAM). The advantage of M-PSK and M-ary QAM data signals is that more than one bit can be encoded per symbol ($\log_2 M = k_M$ bits per symbol). Thus, the spectral efficiency can be enhanced. The bit rate R_B can be calculated from the symbol rate R_{Sy} with $R_B = k_M R_{Sy}$.

The total bit rate of a data-stream can be increased if a higher number of data signals are transmitted simultaneously. CWDM, DWDM or OFDM [16] *multiplexing techniques* efficiently increase the utilization of the about 60 THz fiber bandwidth (from 250 THz (1.3 µm) to 190 THz (1.5 µm)), see *Appendix A.4*.

While much of the development work in recent years was focused on the implementation of high-capacity backbone networks, the speed of metro network and access networks are still lagging behind. Legacy metro networks are typically based on circuit-switched synchronous optical network/synchronous digital hierarchy (SONET/SDH) technology. Legacy SONET/SDH metro networks carry bursty data traffic relatively inefficient, resulting in a bandwidth bottleneck at the metro level. This bandwidth bottleneck, which is also known as metro gap, prevents high-speed clients and service providers in local access networks from tapping into the vast amounts of bandwidth available in the backbone networks [4]. At present, there is a strong worldwide push toward bringing fiber closer to individual homes and businesses with the goal to alleviate the last mile bottleneck. Using intelligent high-speed optical components in the metro and access networks will help to increase the bandwidth. For example, optical processing of all kind of data signals requires *all-optical switching* with OXC between the metro and access rings. These OXC have to offer functionalities such as grooming, all-optical multi-wavelength 3R regeneration (re-amplification, re-shaping and re-timing) and wavelength conversion of highest speed data signals. The immense traffic growth causes also the need of active elements such as *ROADM in the access networks* which comprise nowadays only passive elements if operated optically.

The strong growth of mobile and wireless traffic requires an *optical network topology* that, including network elements, *supports various kinds of data signals*. An overlay of signals is expected that encode information in the optical field, in the intensity and also radio-signals modulated onto an optical carrier. The latter signals are called radio-over-fiber (RoF) signals. Additionally, the increasing number of network elements causes a loss budget issue. Here, *reach-extending and loss compensating optical amplifiers* have to be used to overcome this bottleneck.

1.4 Trend towards Next–Generation All–Optical Networks

A next-generation all-optical network scenario, which could cope with the traffic demand of the clients in the year 2015 and beyond, is schematically shown in Fig. 1.5.

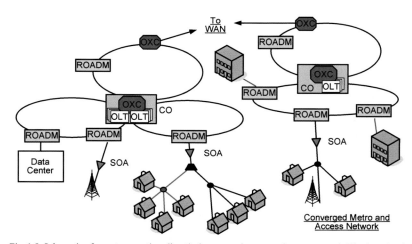

Fig. 1.5: Schematic of a next-generation all-optical converged access and metro network. The data signals on different wavelength channels from the WAN (not shown here) are switched by means of OXC to the metro-core rings. At the edge to the metro-access ring centralized CO are located. In the CO several OLT send the data signals to the access ring towards the end user. Highly flexible, high-speed and long-reach networks are supported. SOA are used in the ROADM and the access networks for amplification and in the OXC for nonlinear signal processing. In the access area only the SOA for the downstream path are shown.

The network connects the user premises equipment all-optically with the high-speed optical backbone network switches. In the WAN (not shown in Fig. 1.5), a large number of wavelengths channels on a DWDM grid are transported over a single fiber. It is predicted that a single wavelength channel carries data signals with bit rates exceeding 100 Gbit/s. This will be achieved with advanced modulation formats, e.g., a 25 GHz DWDM grid with dual-polarization data encoding and a 12.5 GBd 16QAM modulation. Switching and data aggregation are performed in the optical domain with OXC at highest speeds at the edge of the WAN to the metro network.

The metro-core ring network carries several wavelength channels with a bit rate of around 40 Gbit/s...100 Gbit/s. Here, beside advanced formats, high-speed OOK data signals comprising optical-time-division multiplexing (OTDM) tributaries may also be used. At the edge between metro core and metro access networks, main CO are located. In each CO, OXC performs switching and traffic aggregation in the optical domain. Additionally, optical wavelength converters integrated in the OXC avoid wavelength blocking issues. These wavelength converters may rely on SOA and filtering techniques. Several OLT launch data signals on different wavelengths into the metro-access ring towards the end user equipment. The access networks offer a long reach and a high flexibility and scalability in terms of bit rate and modulation formats. Since these next-generation access networks are linked by the metro-access rings via a centralization of CO (compare Fig. 1.5 with Fig. 1.4), this topology is called converged metro and access network. The access network enables high-speed connectivity for each user of around 1...10 Gbit/s, flexible bandwidth allocation by ROADM, long-reach of about 100 km in a TDM/WDM approach, a high split ratio (e.g., 1:512), and a flexible connectivity. This way enterprises, data centers or universities can also be connected directly by an access ring via the ROADM. To increase the bandwidth of this so-called optical trails advanced optical modulation formats may be used. SOA are integrated within the ROADM and before the WDM de/multiplexer in the access networks to increase the overall network loss budget, see Fig. 1.5. Mobile users can connect to these high-speed networks via antennas located in the access networks and thus contribute with RoF data signals.

It should be mentioned that dispersion-compensated fiber or zero-dispersion fiber, digital signal processing and forward error correction will be used in such networks to enable a long reach at these high bit rates.

1.4.1 SOA–based Access and Metro Network Convergence

In the last sections, the trend towards high speed, flexible, versatile and future-proof all-optical networks have been presented. The real benefit from using an all-optical access network approach compared to the latest-generation VDSL is the available broadband optical spectrum of the fiber (large number of WDM channels) and the availability of SOA which offer gain over the entire wavelength region. If a data signal stays in the optical domain all the way down to the user premise, it will experience a huge attenuation due to the large number of network elements such as OXC, ROADM, wavelength division multiplexers, and power splitters. Thus, loss-compensating optical amplifiers and reach extenders are required to increase the tolerable loss budget. Besides optical amplification, additional functionalities such as wavelength conversion, regeneration, clock recovery, and all-optical switching are required to guarantee that the data signals can be transmitted through the network without OEO conversion.

Therefore, linear optical amplifiers and nonlinear signal processing devices are required in next-generation all-optical networks. In this thesis, SOA are investigated for linear and nonlinear applications.

Fig. 1.6 shows a detailed view of the schematic of the all-optical network scenario introduced in Fig. 1.5. A data signal in the metro-core ring on a wavelength λ_1 (black) is launched into an OXC located at the edge to a metro-access ring, see Fig. 1.6(a). The OXC offers functionalities such as grooming, regeneration, time-slot interchanging of OTDM time slots, OTDM-to-WDM conversion and WDM-to-OTDM conversion. The realization of such an all-optical grooming switch will be presented in *Chapter 6*.

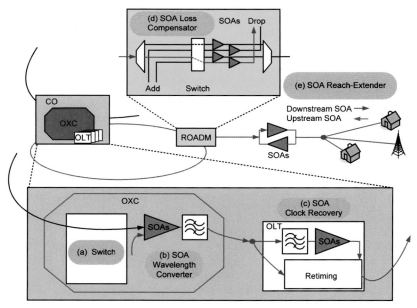

Fig. 1.6: Tasks of this thesis and detailed view of the all-optical network scenario introduced in Fig. 1.5. In *Chapter 4...6* realizations of the individual tasks (a)-(e) are presented. (a) An all-optical grooming switch (OXC) directs a data signal from the metro-core ring to the metro-access ring via an OLT. In the OXC, an SOA-based wavelength converter (b) and in the OLT, an SOA-based clock recovery unit (c) is used. In the metro-access ring, the signal is transmitted through a chain of ROADM which contain loss-compensating SOA (d). If the data signal is dropped by the ROADM to the TDM/WDM access network, SOA increase the reach and the split ratio (e). Especially in the upstream direction, a large input power dynamic range (IPDR) of the SOA is required. This is due to strong power variations at the SOA input introduced by the distance variation of the customer locations.

If the data signal is switched from the metro-core ring by means of the OXC to the metro-access ring, the data signal can be wavelength converted by a nonlinear SOA to a wavelength λ_2 (blue), see Fig. 1.6(b). In this work, quantum-dot (QD) SOA are investigated for wavelength conversion. The achievable bit rate and the optimum wavelength conversion filtering schemes are of interest. The experimental realization of the wavelength conversion schemes are shown in *Chapter 5, Section 5.3*.

After wavelength conversion, the data signal is launched to the OLT. Before the OLT can transmit the data signal to the metro-access ring, a clock recovery is required to re-time the incoming data signal to the local clock of the metro-access network, see Fig. 1.6(c). In this work, an all-optical clock recovery (CR) circuit based on a QD SOA is investigated. The all-optical multi-wavelength CR performance of a Fabry-Pérot filter followed by a QD SOA is demonstrated in *Chapter 5, Section 5.4*.

If the data signal is transmitted in the metro-access ring, it may pass through a chain of ROADM, in which amplification takes place. Here, the question is if a cascade of QD SOA (located in the ROADM, see Fig. 1.6(d)) can amplify OOK and advanced modulation format data signals. Results of a re-circulating loop experiment are presented in *Chapter 4, Section 4.6.3*.

If a data signal is dropped by means of a ROADM towards the end user location, reach-extending QD SOA can be used to extend the reach and the split ratio of the access network, see Fig. 1.6(e). Here, the question in the downstream direction is if the QD SOA are able to amplify advanced optical modulation formats and what the influence of the SOA amplitude-to-phase coupling on the quality of these signals is. In the upstream direction, the SOA input power dynamic range (IPDR) is of particular interest. There a large IPDR of the SOA is required. This is due to strong power variations at the SOA input which are introduced by the distance variation between the customer locations. In *Chapter 4, Section 4.3-4.4*, the IPDR is introduced, design guidelines are provided to optimize the SOA reach extender and measurement results for OOK and advanced modulation formats are presented.

1.5 Conclusion

In conclusion, next-generation all-optical networks require linear and nonlinear SOA, which support besides a large optical bandwidth exceeding at least 60 nm, small footprint, low cost, and high energy efficiency, the following functionalities:

- Amplification of high-speed intensity as well as field modulated data signals
- Transparently cope with different multiplexing techniques (WDM, OFDM)
- Large input power dynamic range and a high burst mode tolerance
- Capability to enable an SOA cascade for various modulation formats and multiplexing techniques
- Wavelength conversion at highest speeds
- Clock recovery at highest speeds

2 Semiconductor Optical Amplifiers (SOA)

Linear and nonlinear semiconductor optical amplifiers (SOA) have attracted much interest in the last few years due to their ability to amplify signals in 10 THz windows over the whole spectral range from 1250 nm up to 1600 nm at reasonable cost [17] and due to their signal processing capabilities [18].

Optical fiber communications systems, especially in the metro and access networks, take advantage of semiconductor-based optical amplifiers and signal processors because of their compact size and high efficiency. These days SOA show promise as in-line amplifiers (reach extender) in fiber-to-the-home (FTTH) applications [17], in photonic integrated circuits (PIC) [19] to boost a data signal from a directly modulated laser source or to compensate for losses in optical switches. SOA are also used for nonlinear signal processing such as wavelength conversion, regeneration, optical switching [20] to name but a few applications.

Depending on the desired application, SOA can offer fiber-to-fiber gain G_{ff} exceeding 25 dB [21]. Today, conventional SOA with a polarization dependent gain (PDG) of down to 0.2 dB are commercially available [22]. Low PDG is important in many cases – for example within a fiber communications network the state of polarization of the data signal typically is unknown. An unavoidable effect which occurs during the data signal amplification is that noise is added to the signal. The corresponding parameter is the fiber-to-fiber noise figure NF_{ff} which typically has values between 5...8 dB [23]. Additionally, in networks in which multiple data signals are transmitted on different wavelengths, an SOA needs to be able to amplify these data signals simultaneously. Here, the SOA gain bandwidth B_G and the saturation input power determine the maximum number of channels which can be handled. An SOA gain bandwidth of up to 100 nm [24] and high saturation input powers $P_{sat}^{in,f}$ exceeding 10 dBm [25] have been demonstrated.

In *Section 2.1*, SOA basics are discussed. The concept of an SOA including working principles, material systems, structures and their growth, is reported. The use of SOA in PIC and their classification as booster amplifier, in-line amplifier and pre-amplifier devices is addressed together with their use as wavelength converters in nonlinear applications.

The content of this section has been published by the author in [B1]. Minor changes have been done to adjust the notations of variables and figure positions. The *Subsection 2.1.6* has been extended with material describing the application of SOA as nonlinear elements.

In *Section 2.2*, all key parameters that characterize the performance of an SOA are discussed in more detail. A simple but efficient model is used to determine the gain G, the noise figure NF, the saturation power P_s, the SOA dynamics (covering inter-band as well as intra-band effects) and finally, the time-dependent effective alpha factor α_{eff}.

The content of this section has been published by the author in [B1] and [J10]. Minor changes have been done to adjust the notations of variables and figure positions. The

Subsection 2.2.7 has been added containing material about the modulation response of an SOA. The *Subsection 2.2.8* has been added describing SOA nonlinearities.

In *Section 2.3*, an excursus on QD SOA is provided, because some features of these devices cannot be explained with the theory of conventional SOA. Advantages and disadvantages of these novel devices are discussed and two operating regimes, a high-carrier injection and a low carrier injection regime, are introduced.

Part of the content of this section (*Subsections 2.3.2* and *2.3.3*) has been published by the author in [J7] and [J10]. Minor changes have been done to adjust the notations of variables and figure positions.

2.1 SOA Basics

In this section, the SOA operating principle is introduced. The condition of achieving optical gain, the more common compound semiconductor heterostructures, and the active-region materials to grow such SOA are discussed, along with the SOA device structure and design. The packaging of SOA into 14-pin butterfly cases and their linear and nonlinear applications are outlined.

2.1.1 Absorption and Emission of Light

An SOA is a semiconductor waveguide with a gain medium. In an SOA the gain is obtained by injecting carriers from a current source into the active region. These injected carriers occupy energy states in the conduction band of the active material leaving holes in the valence band. Electrons and holes recombine either non-radiatively or radiatively, in this case releasing the recombination energy in form of a photon. Three radiative processes are important in such structures, namely induced (= stimulated) absorption, spontaneous emission and induced (= stimulated) emission of photons.

First these processes are discussed for a two-level system and subsequently the situation for a semiconductor material is illustrated. The three radiative types of transitions a microsystem can undergo are explained schematically in Fig. 2.1.

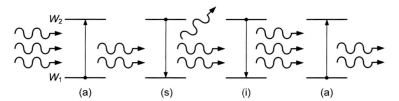

Fig. 2.1: Interaction of a two-level microsystem with electromagnetic radiation, photon energy $hf_s = W_2 - W_1$. A sequence of processes is shown with (a) absorption, (s) spontaneous emission, (i) induced (= stimulated) emission, and (a) absorption of photons, [B1].

Absorption. A microsystem in its ground state W_1 can absorb radiation at a frequency $f_s = (W_2 - W_1) / h$ (photon energy hf_s, Planck's constant h) and make an upward transition to its higher energy level W_2. This absorption process is obviously induced or stimulated by an existing electromagnetic wave. The absorption rate depends on the electromagnetic energy density, and on the number of microsystems in the ground state, Fig. 2.1(a).

Spontaneous Emission. An excited microsystem in energy level W_2 can make a downward transition to the ground state W_1 "spontaneously" (apparently without any interaction with other photons) by emitting a photon with energy $hf_s = W_2 - W_1$, Fig. 2.1(s). The spontaneous emission rate depends on the number of excited microsystems. The spontaneous emission can be regarded as being induced by so-called zero-point fluctuations. This energy fluctuation around the electromagnetic field expectation zero represents a perturbation for an excited microsystem and may therefore induce random transitions to the ground state. The spontaneously emitted photons will be found with equal probability in any possible mode of the electromagnetic field.

Induced Emission. A microsystem in an excited level W_2 can also make a downward transition to the ground state W_1 in the presence and induced by an external radiation (incident photon) of frequency $f_s = (W_2 - W_1) / h$. As in the case of (induced) absorption, the emission rate depends on the electromagnetic energy density and on the number of microsystems in the excited state, Fig. 2.1(i). In contrast to the spontaneous emission process the emitted radiation is in all respects coherent to the stimulating radiation. Therefore, the induced radiation adds with the same polarization and phase to the stimulating field which thus becomes amplified.

For a microsystem in thermal equilibrium the occupation probability $w_p(W_i)$ of the various energy levels W_i at any temperature T is given by the Maxwell-Boltzmann statistics (degeneracy g_{di} of level W_i, Boltzmann's constant $k = 1.380658 \times 10^{-23}$ Ws/K)

$$w_p (W_i) = \frac{g_{di} \exp\left[-W_i / (kT)\right]}{\sum_i g_{di} \exp\left[-W_i / (kT)\right]}. \tag{2.1}$$

For non-degenerate ($g_{di} = 1$) two-level microsystems as in Fig. 2.1, the relative occupation probability numbers $N_{1,2}$ of microsystems with energy states $W_{1,2}$ in thermal equilibrium can be derived from (2.1)

$$\frac{N_2}{N_1} = \exp\left[\frac{-(W_2 - W_1)}{(kT)}\right], \qquad N = N_1 + N_2. \tag{2.2}$$

The quantity N is the total number of microsystems. As seen from (2.2), in thermal equilibrium the excited state W_2 is less densely populated than the ground state W_1. With induced absorption as described in Fig. 2.1(a), the population number N_2 may be increased in proportion to the photon number N_p, and in proportion to the time t. Spontaneous emission reduces N_2 in proportion to t, and stimulated emission diminishes N_2 in proportion to N_p, and t. Therefore, in the presence of an electromagnetic field of photon energy hf_s, a dynamic equilibrium will be reached for which the number of spontaneous and induced emission processes equals the number of stimulated absorption processes. If N_p is so large that spontaneous emission may be neglected, a state of dynamic equilibrium with $N_2 = N_1$

(including spontaneous emission $N_2 \leq N_1$ holds) may be reached so that the number of stimulated emission processes equals the number of stimulated absorption processes. The medium is called *transparent* in this case. However, with a two-level system it is impossible to achieve what is called "population inversion" with $N_2 > N_1$, where the number of stimulated emission processes is larger than the number of absorption processes. Exactly this case would lead to a net amplification of an electromagnetic wave propagating in such a medium.

Semiconductors

The situation is significantly different with a semiconductor material. In the following, it is now shown that population inversion and therefore a net medium gain can be obtained.

The valance band (VB) and conduction band (CB) states of the semiconductor can be associated with multiple closely spaced energy levels centered around W_1 and W_2 in a low and in a high energy state, respectively. According to the equilibrium distribution (2.1), the occupation probability of the lowest energy sublevel (W_0) is highest, and of the highest energy sublevel (W_3) is lowest, so that the absorption from the lowest energy states to the highest ones is a most probable process, see Fig. 2.2. If an electron is excited to a W_3 sublevel, it would relax quickly within the band to the lower energy sublevel W_2' (in technical terms one would say that the intra-band carrier relaxation time is short). This way all excitations to a level W_3 will almost instantaneously be transferred to level W_2' which becomes more densely populated while level W_3 stays lowly populated.

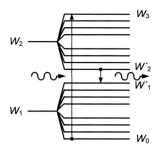

Fig. 2.2: Pump mechanism using energy levels inside the energy level group (pseudo-four-level system), [B1].

Emission from a strongly populated sublevel (W_2') to a sparsely populated sublevel (W_1') is very probable. Therefore, the maximum for absorption is found at higher frequencies (shorter wavelength) than the maximum for spontaneous emission.

This way, pump light with photon energy $hf_s^{(a)} = W_3 - W_0$ can be absorbed thereby generating electron-hole pairs. If the pumping is strong enough, then population inversion can be achieved. Optical pumping is not the only way to arrive at population inversion. It is also possible, and indeed preferable, by injecting electrons and holes electronically into a forward biased semiconductor pn-junction. For a hypothetical temperature $T = 0$ K the "pump energy" $hf_s^{(a)} = eU$ provided by the forward voltage U (elementary charge $e = 1.602 \times 10^{-19}$ C) would define the energetic difference at which electrons and holes could be injected, i.e., the difference of the quasi-Fermi levels $W_{Fn} - W_{Fp} = hf_s^{(a)} = eU$ for electrons in the CB (W_{Fn}) and for holes in the VB (W_{Fp}), respectively. The energy $hf_s^{(e)} = W_2' - W_1'$ of the emitted photons is therefore smaller than $W_{Fn} - W_{Fp}$ but necessarily larger than the bandgap W_G [26], so that the general inversion condition for amplification of an electromagnetic wave by a semiconductor is determined by

$$W_G < hf_s^{(e)} \leq W_{Fn} - W_{Fp} . \tag{2.3}$$

2.1.2 Compound Semiconductors and Heterostructures

In the previous section the general material properties for generation and amplification of light have been discussed. Such properties are, e.g., found in the III-V compound semiconductor (In,Ga)(As,P), a material system which will be discussed here.

In this material system the bandgap W_G and hence the bandgap wavelength λ_G and the refractive index n depend on the composition. In all practical instances the compounds have radiative direct band-to-band transitions without the need of phonon interactions. Further, the lattice constant of the compound may be chosen to match the lattice constant of a binary substrate semiconductor. Lattice matching is very important for several reasons:

- A close lattice match is necessary in order to grow high-quality crystal structures.
- Excess lattice mismatch between the heterostructure layers (adjacent semiconductor layers with different W_G) would result in crystal imperfections which lead to nonradiative recombinations.
- A moderate lattice mismatch (including either tensile or compressive strain) enables bandgap engineering and parameter optimization, as is outlined in more detail below.

Semiconductor	W_G / eV ($\lambda_G / \mu m$)	n at λ_G	a_{La} / nm
GaAs, direct	1.424 (0.871)	3.655	0.5653
InAs, direct	0.36 (3.444)	3.52	0.6058
InP, direct	1.35 (0.918)	3.45	0.5869
GaP, indirect	2.261 (0.548)	3.452	0.5451
(In$_{1-x}$Ga$_x$)(As$_y$P$_{1-y}$) latticed-matched to InP direct: $y \leq 1$ $x = y/(2.2091 - 0.06864\,y)$	$1.35 - 0.72y + 0.12\,y^2$ $1.35...0.75$ (0.918...1.653)	$3.45 + 0.256y - 0.095\,y^2$ $3.45...3.61$	0.5869

Table 2.1: Material system (In$_{1-x}$Ga$_x$)(As$_y$P$_{1-y}$). W_G bandgap, λ_G bandgap wavelength, n refractive index, a_{La} lattice constant, [B1].

By varying the compound's composition, the bandgap as well as the refractive index and the lattice constant can be engineered. SOA realized with the (In$_{1-x}$Ga$_x$)(As$_y$P$_{1-y}$) material system operate in the wavelength range of $\lambda = 0.92...1.65\ \mu m$ and are typically grown on lattice-matched indium phosphide (InP) substrates. Alternatively, gallium arsenide (GaAs) substrates may be chosen which provide amplification in the wavelength region from $\lambda = 0.87\ \mu m$ (GaAs) down to $\lambda = 0.68\ \mu m$ ($In_{0.49}Ga_{0.51}P$). Table 2.1 provides a summary [27], [28].

Heterostructures
SOA typically consist of heterojunction structures [29] composed of doped semiconductors with different bandgap energies W_G. When an impurity atom is implanted into a crystal, its perfect periodicity is destroyed and additional energy levels for electrons located near the band edges are created. These levels are either near the conduction-band edge for donating an electron to the conduction band (donator, n-doping), or they are near the valence-band edge for accepting an electron from the valence band (acceptor, p-doping), see the position of the Fermi-energy W_F in Fig. 2.3.

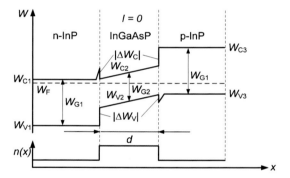

Fig. 2.3: Energy-band diagram of an InGaAsP/InP double-heterostructure in thermal equilibrium and the schematic of refractive index n as a function of the growth direction x, [B1].

An energy-band diagram of a typical heterostructure as a function of the layer growth direction x is shown in Fig. 2.3 for thermal equilibrium. The active InGaAsP region thickness, d, is typically in the range $0.1...0.5\ \mu m$. The active layer is not intentionally doped. However, due to diffusion of p-dopants it typically is slightly p-doped. The neighboring layers consist of InP which has a larger bandgap $W_G = W_C - W_V$ (with the CB edge energy W_C and the VB edge energy W_V) than InGaAsP and therefore comes with a lower refractive index n, see Fig. 2.3. The band-edge energies (and therefore the carrier concentrations) are not continuous but exhibits steps by $|\Delta W_C|$, $|\Delta W_V|$ which lead to spikes and jumps in the band diagram.
In detail, this heterostructure exhibits the following features:
Potential wells: The low bandgap of the active medium creates a potential well for carriers which become confined to the active region. If the difference of the quasi-Fermi levels exceeds the bandgap, $W_{Fn} - W_{Fp} > W_G$, population inversion occurs and optical amplification results.

Larger W_G in the InP cladding layers: The larger bandgap in the InP regions prevents re-absorption in the non-inverted cladding regions.

Smaller n in the InP cladding layers: A structure with a larger refractive index in the core which is clad by a lower refractive index material forms a slab waveguide that provides guiding in the vertical direction. Lateral guiding results from the lateral gain profile, or by additionally adding low-index regions laterally.

For the 3-layer structure presented in Fig. 2.3 both the carrier and the field confinement are determined by the thickness d of the active layer and by the bandgap energy W_G.

Fig. 2.4 shows the energy-band diagram of the double-heterojunction of Fig. 2.3 for the case of a large forward current so that the flat-band case is approximately reached. Far away from the junction the quasi-Fermi levels of electrons and holes are practically identical. However, inside the thin active p-InGaAsP we have $W_{Fn} > W_{Fp}$ due to the carrier injection. The electrons and holes are confined to the potential well inside the active p-InGaAsP layer. Due to current injection, the semiconductor may be population-inverted and the quasi-Fermi level for electrons W_{Fn} shifted towards or beyond the CB, while the quasi-Fermi level for holes W_{Fp} is shifted towards or beyond the VB edge.

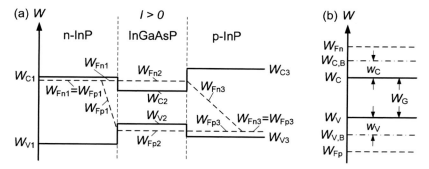

Fig. 2.4: Energy-band diagram of a double-heterojunction with a forward bias voltage U ($T = 0$ K and flat-band case assumed) in (a). The quantities w_C and w_V are energies measured from the respective band edges into the bands which is shown in (b), [B1].

Physics of Gain – Material Gain

The material gain g_m is equivalent to the fractional increase of photons per unit length. It can be approximately calculated by Fermi's golden rule, a quantum mechanical expression for the transition probability of electrons from the conduction to the valence band [20], [30]-[33]

$$g_m \sim \int_0^\infty |M_{ave}|^2 \, \rho(k_{tr}) \big[f_C(w_V(k_{tr})) + f_V(w_C(k_{tr})) - 1 \big] \, \Im(hf_s - w_{CV}(k_{tr})) \, dk_{tr} \, . \qquad (2.4)$$

The terms in the integral in (2.4) are:

- k_{tr} is the electron-hole transition wave vector.
- $|M_{ave}|^2$ is the squared dipole matrix element describing the quantum-mechanical properties of the inter-band transition.

- The number of available states per energy interval to be occupied at a certain energy level in the semiconductor material is described by the density of states (DOS). The DOS in the k_{tr}-space is $\rho(k_{tr})$.

- The occupation probabilities for CB electrons and VB holes under non-equilibrium conditions (forward bias supplied to pn-junction) are described by the Fermi functions of the CB f_C or VB f_V, respectively.

- $\Im(hf_s - w_{CV}(k_{tr}))$ is the linewidth-broadening function that takes into account the finite lifetime of the carriers due to scattering effects. The photon energy is hf_s and $w_{CV} = w_C + W_G + w_V$ is the transition energy from the CB to the VB with the bandgap energy W_G and the band energies of the CB w_C and VB w_V, respectively.

Details of the calculation of g_m can be found in [32]. Fig. 2.5 shows the spectral dependance of the material gain for InGaAsP [33]. The gain region occurs at 1.5 μm. The curves have been calculated for carrier densities ranging from $N = 0.5 \times 10^{18}$ cm^{-3} to 3.0×10^{18} cm^{-3} in steps of 0.5×10^{18} cm^{-3}. It is assumed that the electron- and hole densities are equal, i.e., that $N = P$. Obviously, the material gain increases with increasing carrier injection, and the maximum of the material gain shifts to lower wavelengths basically due to band-filling effects. In all practical instances the material gain is not directly calculated by (2.4). One normally goes back to parameterized material gain curves that have been derived from experiments [32].

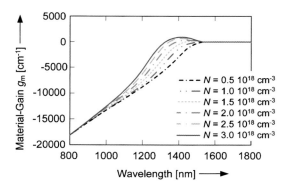

Fig. 2.5: Calculated material gain g_m for different carrier densities for 1.55 μm bandgap InGaAsP. The picture is taken from [33] and [B1].

2.1.3 Properties of the Active Region

Semiconductor optical amplifiers require an active region consisting of a direct bandgap material which gives a high probability for radiative recombinations. SOA can differ in the dimensionality of the electronic system of their active region. The classification is done in terms of the mean free path length l_{fp} of the carriers in the active region with respect to the de Broglie wavelength λ_{dBr} of the carriers.

In this sense, if the carrier motion is not restricted, we talk about bulk SOA. Fig. 2.6(a) shows the DOS as a function of energy W for such a three-dimensional (3D) structure as well as the carrier concentrations of electrons N (blue area under the curve). The DOS shows a square-root dependency on the energy.

If the carrier motion is limited to a layer of a thickness d, which is in the order of the de Broglie wavelength (10...100 nm), discrete energy levels occur for this direction [34]. If the carrier motion is restricted to two dimensions (2D), it is common to talk about a quantum "well" SOA (QW, strictly speaking, it is a quantum film or quantum layer). The DOS corresponding to such discrete energy states is constant as depicted in Fig. 2.6(b).

If the carrier motion is restricted to one dimension (1D), a quantum wire or quantum dash (QDash) active region results. The quantum wire DOS is proportional to $1/\sqrt{W}$ as shown in Fig. 2.6(c).

Finally, if the carrier motion is fully restricted (zero dimension 0D), we talk of a quantum-dot (QD) SOA. The atom-like DOS for a QD is expressed by a delta function $\delta(W)$ as depicted in Fig. 2.6(d).

The advantages in reducing the dimensionality of the electronic systems are manifold. In particular, the need of a temperature independent material gain as well as a reduction in the operating current density for a certain material gain drives these approaches.

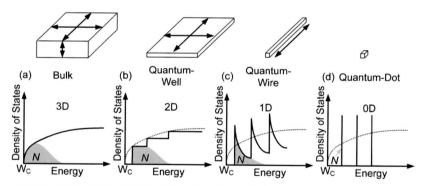

Fig. 2.6: Density of states (DOS) for SOA. (a) In bulk material, the carriers are not restrained in their movement. However, if the carriers experience a spatially dependent potential, the dimension of which is smaller than the mean free path length of the carriers, then (b) in a quantum well the carrier motion is limited to a plane, (c) in quantum wire (or quantum dash) the carrier motion is limited to one direction, and in a quantum dot the carrier motion is fully restricted in all dimensions. The corresponding electron concentration N (blue area) is shown. W_C is the conduction band edge, [B1].

Bulk SOA: The active region thickness d of a bulk double-heterostructure SOA is typically 100-500 nm. As already mentioned, a common compound used to fabricate such bulk SOA structures for the wavelength region of 1.3 µm or 1.55 µm is based on an $In_{1-x}Ga_xAs_yP_{1-y}$ active layer with adjacent InP layers grown on an InP substrate. The gain bandwidth is determined by the relationship of the material fractions (x and y in Table 2.1). Further, lattice matching is needed to reduce non-radiative recombination at imperfections.

If bulk SOA are fabricated with a slab-like waveguide, strong PDG will be found due to different optical confinement factors (fraction of light in the active region) for the fundamental transverse electric (TE) and transverse magnetic (TM) propagating modes. Since a low PDG is very important due to the unknown state of polarization at the amplifier input, several approaches are in use to reduce PDG. One possibility is the fabrication of virtually ideal square waveguide cross-sections (strip waveguide) to achieve identical confinement factors for the TE and TM mode. However, since such square waveguides are difficult to fabricate, a strained active bulk layer can be grown to adjust the material gains for the TE and TM mode.

Strained layers result from choosing the material composition with a slight mismatch in the lattice constant relative to the substrate material [22]. If properly chosen, the generated difference in the gain for the TE mode and TM mode counteracts the difference in optical confinement for the TE mode and TM mode. In effect, these bulk SOA have virtually polarization independent gain. In [22] it is shown that a tensile strain for the active bulk layer increases the TM material gain relative to the TE material gain, and the PDG is minimized.

Another approach, which is widely used to achieve low PDG, is the growth of a thin layer with higher bandgap material on either side of the active region, a so-called separate confinement heterostructure (SCH) [22]. This layer has a refractive index between that of the active region (InGaAsP) and the cover (InP) or substrate layer (InP). The inclusion of a SCH structure gives control over the confinement factor ratio for the TE and TM mode via its refractive index and thickness, but it also changes the dynamic characteristics of the SOA.

QW SOA: If the carrier motion normal to the active region is restricted by properly choosing the layer thickness d (down to $10\ldots100$ nm), the carriers are confined in a potential "well".

QW SOA or multi quantum-well (MQW) SOA with lattice matching are strongly polarization dependent with high gain for the TE polarization and considerably lower gain for the TM polarization. Typically, polarization insensitive operation has been shown using both compressive and tensile strained wells based on InGaAsP/InP ([35]-[37]), tensile strained QW SOA based on AlGaInAs/InP for the wavelength region of 1.3 µm ([38]-[40]) and MQW SOA tensile strained wells based on AlGaInAs/InP for operation at 1.55 µm [41].

For a reduced temperature dependance of the material gain, AlGaInAs/InP is used instead of InGaAsP/InP. The larger conduction band offset with AlGaInAs results in a higher electron confinement in the QW. Another promising material system is GaInNAs on lattice-matched GaAs substrates. The GaInNAs compound is particularly promising for uncooled amplifier operation because the material's temperature dependance of the gain is superior to that of conventional InP-based structures. At a wavelength of 1.3 µm, QW SOA have been demonstrated with (Al)GaInP or (Al)GaAs cladding layers on GaAs substrates [42], and 1.5 µm MQW SOA were reported with GaInNAs/GaInAs layers on InP substrates [43].

So far, SOA were mostly developed on the InGaAsP/InP or AlGaInAs/InP material systems. However, the overall energy efficiency of these devices is unsatisfactory and becomes worse still at elevated temperatures. One of the principal causes of the poor electro-optical conversion efficiency is that a large fraction of the pump energy is converted into unwanted heat via non-radiative recombinations. It is predicted that new materials such as bismides and diluted nitrides on InP and GaAs offer exciting possibilities to reduce or remove the detrimental effects of, e.g., Auger recombination [44]. It has been further shown for the case of GaAsBi that about 10 % Bi is needed to suppress the problematic Auger process [45], and optically pumped lasing has already been demonstrated [46]. Photoluminescence in the wavelength range used for fiber optic communication and a beneficially weak temperature dependent variation of the energy gap compared to InGaAsP were reported for the quaternary compound $Ga(N_{0.34}Bi_{0.66})_zAs_{1-z}/GaAs$ [47]. These new material systems are currently investigated in a number of research projects to determine to what extent the theoretically predicted benefits can be realised.

QD SOA: QD SOA with InAs QD on InP substrates are fabricated using a self-organized growth technique for operation wavelengths at 1.55 µm [48]-[50]. Also, InAs QD on InGaAs/GaAs layers [51], [52] are available at wavelengths around 1.3 µm. The typical size of these QD is 10 nm in height and 50 nm in width [49]. Usually, several QD layers are vertically stacked and separated with a spacer layer having a thickness exceeding 30 nm [50]. QD shapes are neither cubic nor ball-shaped which results in strong PDG. One approach to counteract this PDG is the fabrication of columnar quantum dots by vertically stacking QD layers with a spacer layer thickness of around 1...3 nm only. This approach has successfully been used to show polarization independent gain in QD SOA [53]-[55].

Several techniques exist to grow SOA. Typically, metal-organic chemical vapor deposition (MOCVD) or molecular beam epitaxy (MBE) are used. Beside the growth of lattice-matched structures, the strained layer epitaxial growth is also used for special purposes as explained previously. Lattice mismatch between the epitaxial layers and the substrate compared to the substrate lattice constant is around 1.5 %.

In contrast to the growth of several nm thick layers on top of a planar substrate for QW SOA, SOA devices with QD active regions are fabricated by a self-organized growth process, where the lateral dimensions of the quantum dots can be controlled simply by choosing the appropriate growth parameters. The Stranski-Krastanov (SK) growth regime, in which self-assembled quantum dots are formed, is based on a slight lattice mismatch between the substrate and the quantum-dot material. In the case of InAs dots on a GaAs substrate, this lattice mismatch is ~ 7 %. At the beginning of the SK self-assembled quantum-dot growth mode, a compressively strained film (wetting layer) is formed. At a critical InAs coverage of around 0.5 nm, the accumulated strain is elastically relieved by the formation of small 3D islands. Beyond the critical InAs thickness, the additional InAs can migrate freely on the sample surface, adding to the 3D island growth. The growth of a capping layer finishes the dot formation. If the capping layer material has a smaller bandwidth than the substrate, a quantum well structure around the dots is formed. This technique is used to extend the emission wavelength of the dots (dot-in-a-well-structure) [56].

QD SOA promise several advantages, such as an ultra-fast QD gain response in the order of 1-25 ps due to a fast QD refilling through the carrier reservoir represented by the wetting layer [J10]. The ideal QD SOA also offers a large gain bandwidth exceeding 120 nm which is due to QD size fluctuations during the growth [24], low temperature dependance of the gain [57], and low chirp even under gain saturation [58]. The enhanced multi-wavelength capability which is expected compared to bulk/QW SOA has up to now been found in simulations only [59], [C39].

2.1.4 SOA Structures and Devices

SOA are fabricated with an active region material (nominally undoped, i.e., intrinsic) with a larger refractive index compared to its n-cladding and p-cladding layer materials. Such a structure forms a pin-junction. In general, the refractive index difference between core and cladding material is around 0.3 for InGaAsP/InP structures. This strong index difference leads to a vertical guiding of the optical mode. However, with a slab waveguide, the light is unconfined in the lateral direction, parallel to the pin-junction plane. This issue can be solved by tailoring the SOA device structure. In Fig. 2.7, various SOA cross-sections with lateral light confinement are presented: (a) stripe-geometry for gain-guided SOA, (b) ridge waveguide for a weakly index-guided SOA, and (c) buried heterostructure for a strongly index-guided SOA.

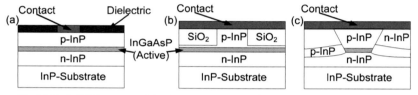

Fig. 2.7: Typical SOA structures: (a) stripe-geometry structure for gain guiding, (b) ridge-waveguide structure for weak index guiding and (c) buried heterostructure for strong index guiding. The figures are modified from [61], and not drawn to scale, [B1].

The technically simplest and cheapest solution for guiding the light is a gain-guided structure which limits the current injection to a narrow region [60]. This stripe-geometry (Fig. 2.7(a)) is easy to fabricate since a contact region is required only after the growth process of the n-cladding layer, the active region and the p-cladding layer. The contact consists of a dielectric layer which has an opening of 4…6 μm through which the current is injected via a metal contact. In the direction parallel to the pin-junction plane, the lateral spread of injected carriers introduces a lateral gain profile. Since the active region has large absorption losses beyond the central stripe, the light is confined to the stripe region which shows gain (active region). As the light-confinement is influenced by the injected carriers, a so-called gain-guiding structure is fabricated in which the mode is weakly guided [61]. The counteracting antiguiding due to the higher carrier concentration below the contact stripe and its associated reduction of the refractive index usually has a lesser effect.

In order to improve the performance, stronger waveguiding is achieved with so-called index-guided structures. Here, an index step is introduced also in the lateral direction, i.e., parallel to the pin-junction plane. A ridge waveguide (Fig. 2.7(b)) is comparatively simple to fabricate. After the growth of the n-cladding, the active region and the p-cladding a ridge is formed by etching parts of the *p*-layer. The ridge width (waveguide width *w*) is typically 2...4 µm. Next, a dielectric layer (SiO_2 or Si_3N_4) is deposited to block the current flow and to introduce a refractive index difference of around 2 between the semiconductor material and the dielectric material. In this way, the mode is confined to the ridge waveguide by weak index-guiding (effective transverse index difference of 0.03). Finally, the metal contact layer is fabricated.

Strong index-guiding SOA have a buried heterostructure (Fig. 2.7(c)). Here, the InGaAsP active layer is "buried" by surrounding it on all sides by InP regions which have a lower refractive index [31], [61]. The layers are fabricated such that any current is efficiently blocked between the *n*-InP/*p*-InP, so that a good current confinement is guaranteed. The effective transverse index difference between the core and cladding regions is around 0.3. However, the fabrication of buried-heterostructure SOA is complex, and several growth and etching steps are required to form the current blocking layers and the contacts.

After the definition of a gain-guided or an index-guided structure, the SOA chip is fabricated. Normally, SOA devices are fabricated as traveling-wave amplifiers. In ideal traveling wave amplifiers (no facet reflectivities as opposed to resonant Fabry-Pérot amplifiers), the optical wave makes only a single pass through the active layer. Typically, an SOA has a length of roughly 0.5 to 4 mm. Antireflection multi-layer dielectric coatings with a thickness of approximately a quarter of a wavelength reduce the total facet reflectivity down to around 10^{-4}. However, residual facet reflectivities are always present, so the SOA becomes a laser oscillator if the gain is sufficiently high. To further suppress lasing, the waveguide is often tilted by 7...10° with respect to the facet normal, thereby reducing the residual reflection from the end facets even more.

Besides preventing lasing, a low facet reflectivity is also desirable to avoid periodic variations of the gain with the cavity mode spacing, so-called gain ripple. The larger the reflectivity of the input and output facets the larger the gain ripple is. The use of antireflection coated facets as well as tilted waveguides lead to an SOA with gain ripple below 0.2 dB.

Typically, SOA chips are mounted on a heat sink to efficiently remove the dissipated heat. Further, single-mode operation of the SOA is desirable which sets limits for the maximum waveguide height *d*, the waveguide width *w* and the maximum refractive index difference between core and cladding material, respectively.

2.1.5 Packaging and Photonic Integrated Circuits

SOA chips are typically packaged into hermetic 14-pin butterfly cases which are free of organic materials (Fig. 2.8(a)). This technology assures high operation stability, long term reliability, low power consumption, and supports operating temperatures in the range – 30...+ 60°C. SOA packaging follows similar rules as the packaging of laser diodes. A lensed input fiber is mechanically stabilized and fixed to the input facet, Fig. 2.8(b). The SOA chip is temperature stabilized with a thermoelectric cooler (TEC) to keep the chip at the operation temperature. The temperature is measured using a thermistor located close to the sample. The contact pads of the pin-junction are wire-bonded to the butterfly pads of the case for current supply. The lensed output fiber is also mechanically stabilized and permanently fixed.

Today, SOA are also often found as subcomponents of PIC. A PIC is a highly integrated device performing multiple functions on a single chip. PIC are gaining interest due to the small footprint, the low power consumption and potentially due to lower cost compared to discrete solutions. PIC are built from substrate materials such as InP, GaAs, lithium niobate ($LiNbO_3$), silicon, and glass. To guarantee cost-effective chip integration, the performance of PIC must be competitive with discrete solutions and the manufacturing yield of each element must be very high.

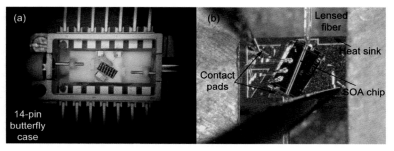

Fig. 2.8: SOA packaging and fiber-to-chip coupling. (a) Top view of an SOA package using typical butterfly cases. (b) Close up of the top view of an SOA chip with fiber-to-chip coupling, [B1].

Although, the InP system is not inexpensive, it reliably integrates active and passive optical devices using standard batch-semiconductor manufacturing processes. InP supports key opto-electronic functions such as light generation, amplification, modulation and detection as well as passive components such as arrayed-waveguide gratings (AWG), filters, and waveguides. To date, the speed of active components is limited due to carrier dynamics in the range of tens to hundreds of ps. Recently, a PIC transmitter with a capacity of 10 × 40 Gbit/s (on-off keying format) has been shown which nicely demonstrates the high integration potential of this material system [19], [62]-[64].

2.1.6 Applications of SOA as Linear and Nonlinear Elements

Linear SOA maintain proportionality between output and input field. They can be used at different locations within a network. In metro networks, linear SOA can be used in optical-electric-optical (OEO) converters to boost the optical signal, or in optical add-drop multiplexers (OADM) to increase the tolerable loss budget. SOA are also good reach extenders in access networks for increasing the distances between the central office and the customer locations, and for providing an increased fan-out (split ratio) to serve a higher number of customers.

Fig. 2.9 shows three common linear SOA applications in networks [62]. A booster amplifier (post-amplifier) increases the power level of a signal at the transmitter prior to transmission. In general, the output power of a laser diode or a tunable laser source is moderate, especially if an external modulator is used. The key feature of the booster is a high saturation output power. Further, the booster should provide bit-pattern effect free amplification of the data signal. In wavelength-division-multiplexing (WDM) systems it should amplify all signals alike across the spectrum. Booster amplifiers normally are polarization sensitive. This is not an issue for boosters as the input signal polarization is known. Booster SOA typically have small-signal fiber-to-fiber (FtF) gains in the order of 10 dB with an in-fiber (power measured in the fiber) saturation output power of more than 10 dBm.

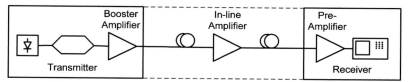

Fig. 2.9: SOA can be used at different positions within a network. An SOA booster raises the power level of the signal at the transmitter, the in-line SOA compensates for fiber losses and the pre-amplifier SOA increases the power level at the receiver to increase the receiver sensitivity, [B1].

An in-line amplifier mainly compensates for fiber losses or splitter losses in an optical transmission system. The most important performance parameters are saturation output power and noise figure because the incoming signals are weak. The polarization dependance of the gain should be as small as possible due to the random state of polarization within a network. Also in-line SOA need to cope with several wavelength channels simultaneously. Further, in-line SOA should process the data signal "transparently" which means that all kinds of modulation formats at any data rate should be amplified without significant degradation. Typical in-fiber saturation output powers are 8 dBm for small-signal FtF gains in the order of 10...20 dB and FtF noise figure between 5...7 dB. In addition, there is an increasing interest in low wall-plug power consumption since in-line amplifiers might be placed outside of network central offices.

A pre-amplifier SOA at the receiver raises the power level of an incoming data signal to enhance the receiver sensitivity. The most important parameters of a pre-amplifier SOA are gain and noise figure as well as low residual polarization dependance of the gain. Typical small-signal FtF gains are between 25...30 dB. The FtF noise figures lie in the region 5...7 dB.

In general, SOA can be designed to operate over a wide range of wavelengths (attempts in the direction of amplification in the range of 1200 nm to 1800 nm with, e.g., QD SOA), and are very compact. They can be modulated very rapidly so they can be used to blank signals when necessary (their gain can be turned on and off in timescales of the order of 1 ns).

Nonlinear SOA do not maintain proportionality between output and input field. They can be used for various applications and at different locations within a network. For example, nonlinear SOA are used in wavelength converters (WC) to transfer the data signal from one wavelength onto a continuous wave (cw) or clock signal at another wavelength [65], [66]. The highest speed wavelength converter achieved until mid of the year 2011 for on-off-keying (OOK) data signals based on bulk SOA performed at bit rates of up to 320 Gbit/s [67]. Recently, a publication on QD SOA showed wavelength conversion at this high bit rate for OOK data signals [68], too.

Nonlinear SOA can also be used to build regenerators which are units to preserve the signal quality within an all-optical network. Since the signal quality degrades during fiber transmission due to chromatic dispersion, noise accumulation and polarization mode dispersion, 2R regenerators which re-amplify and re-shape the optical signals are used to re-establish the signal. In 3R regenerators also a re-timing functionality is included. Fig. 2.10 shows a schematic of a 3R regenerative wavelength converter using SOA. A distorted input signal at a wavelength λ_1 is launched into the SOA-wavelength converter with a 3 R regenerative capability together with a clock signal at a wavelength of λ_2. The ideal power transfer function of the unit is a step-like function representing an optimum optical decision gate for marks and spaces. The output signal at a wavelength of λ_2 shows a very good quality, and the input bit-error ratio is preserved. Some requirements of such 3R regeneration schemes are the suppression of timing jitter and amplitude fluctuations, optical-signal-to-noise ratio (OSNR) and extinction ratio (ER) improvement, a large optical bandwidth and polarization independent operation. Such schemes have been successfully tested for OOK data signals at bit rates of up to 160 Gbit/s [69].

Most wavelength conversion and regeneration schemes which rely on SOA require a filtering scheme [70] to make use of the phase changing capabilities of the SOA. For example a delay interferometer (DI) [66], a pulse reformatting optical filter (PROF) [71], a blue shift optical filter (BSOF) [72] or a red shift optical filter (RSOF) [73], a Mach-Zehnder interferometer (MZI), a Sagnac interferometer (Sagnac) or an ultra-fast-nonlinear interferometer (UNI) [74] may be used.

In recent years, advanced optical modulation formats became a new research field for optical wavelength conversion and regeneration. Here, it needs to be distinguished between phase-insensitive schemes and phase-sensitive schemes. Phase-insensitive schemes make use of a demodulation stage to convert the phase modulation into an intensity modulation with the subsequent use of the already mentioned filtering schemes [75]-[77]. On the contrary, phase-sensitive schemes directly influence the phase and amplitude of an, e.g., phase-sift-keying (PSK) signal. However, it should be mentioned that currently digital signal processing (DSP) at the receiver is a strong competitor to all-optical regeneration schemes.

Up to now, a few nonlinear applications of SOA have been briefly introduced. Other interesting applications using the nonlinearities in SOA are, e.g., clock recovery, optical sampling, demultiplexing, mid-span spectral inversion, pattern recognition, label swapping, XOR and OR gates and optical switches [20].

It needs to be mentioned that the presented nonlinear applications are also achievable using other technologies which are based on the intensity-dependance of the refractive index (Kerr-nonlinearity). Here, e.g., highly-nonlinear fibers (HNLF) [78] and silicon or silicon hybrid waveguides [79] can be used. HLNF-based regenerators for OOK data signals are also available today for multi-wavelength approaches [80].

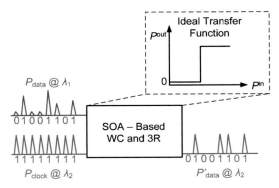

Fig. 2.10: SOA-based wavelength conversion (WC) scheme including re-amplification, re-shaping and re-timing 3R regenerative capabilities. Due to the ideal step-like power transfer function of the unit, the mark and space levels of the distorted input signal can be regenerated.

2.2 Parameters of SOA

The absolute performance of SOA for linear and nonlinear applications is determined by parameters such as the gain, the noise figure, the saturation power, the gain and phase dynamics as well as the chirp described by the so-called alpha factor. In this section, the concepts and the assumptions used to define these parameters are discussed. The description is started with the basic wave equation for an active medium. Then, the gain and the SOA noise figure in direct-detection systems are calculated as well as the SOA saturation power. The carrier dynamics in an SOA due to inter-band and intra-band contributions are described by rate equations and finally the alpha factor relating gain and phase changes is introduced.

2.2.1 Physics of Media with Gain

A signal propagating in a sufficiently pumped SOA is amplified. Starting from the general form of Maxwell's equations [81] a couple of practical assumptions are made to simplify the analysis of light propagation in an SOA:

- The SOA material is non-magnetic, has a sufficiently low free-carrier concentration so that the carriers do not affect the optical properties, is only weakly inhomogeneous and isotropic.
- The medium response to an electric field is described by a medium susceptibility χ.

We start with the description of a homogeneous, weakly amplifying (or attenuating) medium. The complex relative dielectric constant $\bar{\varepsilon}_r$ and the complex refractive index $\bar{n}$ are related by:

$$\bar{\varepsilon}_r(\omega) = \varepsilon_r(\omega) - j\varepsilon_{ri}(\omega), \quad \bar{n}(\omega) = n(\omega) - jn_i(\omega), \quad \bar{\varepsilon}_r(\omega) = \bar{n}^2(\omega), \qquad (2.5)$$

where ε_r, n are the real parts and $-\varepsilon_{ri}$, $-n_i$ are the imaginary parts of relative frequency-dependent dielectric constant and refractive index, respectively [60], [82]. We investigate signal propagation in a sufficiently narrow spectral range, so that $\bar{\varepsilon}_r(\omega)$ and $\bar{n}(\omega)$ can be replaced by their values $\bar{\varepsilon}_r = \bar{\varepsilon}_r(\omega_s)$ and $\bar{n} = \bar{n}(\omega_s)$ at the optical angular carrier frequency $\omega_s = 2\pi f_s$.

The unpumped SOA medium is described by a constant (carrier independent) complex background refractive index $\bar{n}_b = n_b - j\alpha_L/(2k_0)$ with real part n_b and an imaginary part $-\alpha_L/2k_0$ determined by the material loss α_L and the vacuum wave number $k_0 = \omega_s/c_0$ (c_0 speed of light in vacuum). When the medium is pumped, the additional charge carriers (carrier density N) slightly reduce the real part of the background refractive index by n_N and provide gain with a material gain g_m. The complex refractive index due to carrier injection is a perturbation to the complex background refractive index, $\bar{n}_N = -n_N + jg_m/(2k_0)$. Then we can write considering the net material gain $g_{mL} = g_m - \alpha_L$,

$$\bar{\varepsilon}_r = (\bar{n}_b + \bar{n}_N)^2 = \left(n_b - n_N + j\frac{g_m - \alpha_L}{2k_0}\right)^2$$

$$\approx n_b^2 - 2n_b n_N + jn_b\frac{g_m - \alpha_L}{k_0} = n_b^2 - 2n_b n_N + jn_b\frac{g_{mL}}{k_0}. \tag{2.6}$$

With increasing N the carrier-related complex refractive index changes from $\bar{n}_N$ to $\bar{n}_N + \Delta\bar{n}_N \approx -n_N + \Delta n + j(g_m + \Delta g_m)/(2k_0)$, where $\Delta n < 0$ and $\Delta g_m > 0$ hold in this case (with low material losses).

The real and imaginary parts of the refractive index are related by the Kramers-Kronig relation [83]-[85]. This relation is a result of causality in physical systems. Phenomenologically, it has been found that this relation can be simplified for a fixed operating point, and expressed by the linewidth-enhancement factor, also known as the Henry factor or alpha factor α_H [86], [87] which relates the change of the refractive index Δn and the change of the material gain Δg_m (assumption: $\alpha_L \ll g_m$, or $\partial\alpha_L/\partial N = 0$),

$$\alpha_H = \frac{\partial n_N/\partial N}{\partial n_{i,N}/\partial N} \approx -2k_0\frac{\partial n_N/\partial N}{\partial g_{mL}/\partial N} \approx -2k_0\frac{\Delta n}{\Delta g_{mL}} = -2k_0\frac{\Delta n}{\Delta g_m}. \tag{2.7}$$

The alpha factor depends on parameters such as wavelength and current density. Further, the SOA active material strongly influences the alpha factor.

A monochromatic guided wave in a transparent SOA is described by a complex electric field E_y linearly polarized in y-direction, defined parallel to the substrate plane. The field can be written as a product of a transverse modal function $F(x,y)$, an envelope $A(z,t)$, wave propagating terms, and a normalization constant c_p [81],

$$E_y(x,y,z,t) = E_y(x,y)E_y(z,t) = c_p F(x,y)\exp(j\omega_s t)A(z,t)\exp(-jk_0 n_b z). \tag{2.8}$$

In (2.8) and for the following considerations, the background refractive index has to be interpreted as an effective refractive index $n_b = \beta/k_0$ representing the guided-wave propagation constant β. As a next approximation step it is focused on the evolution of $A(z,t)$ only. The term

$c_{\mathrm{p}}F(x,y)$ is fixed such that $|A(z,t)|^2$ equals the total power $P(z,t)$ of the optical field propagating in z-direction.

With the free space impedance $Z_0 = \sqrt{\mu_0/\varepsilon_0}$ (μ_0 is the magnetic permeability and ε_0 the permittivity of free space) we write

$$|A(z,t)|^2 =: P(z,t) = \frac{n_{\mathrm{b}}}{2Z_0} \int \int_{-\infty}^{\infty} |E_y(x,y,z,t)|^2 \, dx \, dy, \quad \frac{|c_{\mathrm{p}}|^2 \, n_{\mathrm{b}}}{2Z_0} \int \int_{-\infty}^{\infty} |F(x,y)|^2 \, dx \, dy = 1. \quad (2.9)$$

In an SOA, the optical mode overlaps only partially with the active volume. The fraction of power propagating in the active region area C (having width w and height d) related to the totally guided power is known as optical confinement factor Γ,

$$\Gamma = \frac{\int_0^w \int_0^d |F(x,y)|^2 \, dx \, dy}{\int \int_{-\infty}^{+\infty} |F(x,y)|^2 \, dx \, dy}. \quad (2.10)$$

Therefore, quantities relevant for the interaction of amplifying medium and optical field have to be modified, leading to an effective net gain g (net modal gain; α_{int} denotes the internal losses with contributions from the active medium α_{L} and the cladding medium α_{Clad}), and an effective real part n_{eff} of the refractive index,

$$\alpha_{\mathrm{int}} = \Gamma \alpha_{\mathrm{L}} + (1-\Gamma)\alpha_{\mathrm{Clad}},$$
$$g_{\mathrm{mL}} \to g = \Gamma g_{\mathrm{mL}} - (1-\Gamma)\alpha_{\mathrm{Clad}} = \Gamma g_{\mathrm{m}} - \Gamma \alpha_{\mathrm{L}} - (1-\Gamma)\alpha_{\mathrm{Clad}} = \Gamma g_{\mathrm{m}} - \alpha_{\mathrm{int}},$$
$$\Delta g_{\mathrm{mL}} \to \Delta g = \Gamma \Delta g_{\mathrm{mL}} - (1-\Gamma)\Delta \alpha_{\mathrm{Clad}} = \Gamma \Delta g_{\mathrm{m}} - \Gamma \Delta \alpha_{\mathrm{L}} - (1-\Gamma)\Delta \alpha_{\mathrm{Clad}} = \Gamma \Delta g_{\mathrm{m}}, \quad (2.11)$$
$$n_{\mathrm{N}} \to n_{\mathrm{eff}} = \Gamma n_{\mathrm{N}},$$
$$\Delta n \to \Delta n_{\mathrm{eff}} = \Gamma \Delta n.$$

Actually, to approximately describe the field in an amplifying guiding medium, the electric field is represented by an equivalent plane wave with amplitude $A(z,t)$ (neglecting the phase front curvature of the actual guided wave) that propagates along the z-direction. With the integration of (2.9) over the cross-section area, the complex electric field $E(z,t) = A(z,t)\exp(j\omega_0 t)\exp(-jk_0 n_{\mathrm{b}} z)$ can be isolated, which is a solution of the wave equation

$$\frac{\partial^2 E(z,t)}{\partial z^2} - \frac{1}{c_0^2}\left(n_{\mathrm{b}}^2 - 2n_{\mathrm{b}}n_{\mathrm{eff}} + jn_{\mathrm{b}}g/k_0\right)\frac{\partial^2 E(z,t)}{\partial t^2} = 0. \quad (2.12)$$

The wave equation (2.12) can be further simplified using the following assumptions:

- The ideal traveling-wave amplifier guides only the fundamental mode in the active region, and the polarization of the field is conserved.
- The bias current supplies carriers, which under all conditions remain uniformly distributed in the active volume.

The envelope $A(z,t)$ of the propagating wave varies slowly in space compared to the wavelength of light (slowly varying envelope approximation) [81],

$$\left| \frac{\partial^2}{\partial z^2} A(z,t) \right| \ll \left| 2k_0 n_b \frac{\partial}{\partial z} A(z,t) \right|. \tag{2.13}$$

If we neglect group-velocity dispersion, because its effect on the optical field envelope is negligible for a typical amplifier length of approximately 1 mm, and then introduce the retarded time $t' = t - z/v_g$ (group velocity v_g) and an amplitude $A'(z,t') = A(z,t)$, a simplified version of the nonlinear Schrödinger equation (NLSE) results [88],

$$\frac{\partial A'(z,t')}{\partial z} = \left(jk_0 n_{eff} + \frac{g}{2} \right) A'(z,t'). \tag{2.14}$$

Eq. (2.14) describes the slowly varying amplitude of the wave during propagation in the active medium of an SOA. In the following we understand that an input amplitude $A'(0,0)$ at $z = 0$ and $t = 0$ leads to a time-delayed output amplitude $A'(L, t - L/v_g)$ at $z = L$. However, because n_{eff} and g are regarded as z-independent global quantities, we drop the retarded-time argument. Any changes of the output amplitude A' are then due to temporal changes $n_{eff}(t)$ and $g(t)$ only. This implies that the SOA length L is much shorter than the length scale, on which $A'(z,t')$ varies. In this spirit, and because we neglect dispersion, we replace in the following $A'(z,t')$ by $A(z)$.

More elaborate SOA models take into account the photon density variations along the SOA waveguide by means of longitudinal segmentation. Thus, the SOA is subdivided along the propagation direction into a number of sections for which the differential equations for carrier density and photon number are solved. The results are evaluated at the boundaries and handed over to the following section [82].

2.2.2 Gain and Phase

In an active material it is essential to learn about the photon density $S(z)$ or the optical power $P(z)$ and the phase $\phi(z)$ at the SOA output. A phasor representation of the envelope $A(z)$ is introduced to separate $\sqrt{P(z)}$ from the phase $\phi(z)$ of the wave,

$$A(z) = \sqrt{P(z)} \exp(j\phi(z)). \tag{2.15}$$

By substituting (2.15) into (2.14) and performing a separation of variables, we obtain differential equations for the power

$$\frac{dP(z)}{dz} = gP(z) \tag{2.16}$$

and for the phase

$$\frac{d\phi(z)}{dz} = k_0 n_{eff}. \tag{2.17}$$

In this simplified model, the net modal gain g is assumed to be independent of the propagation distance z, so that an integration of (2.16) yields an expression for the SOA chip output power $P(L) = P_{out}$ for a propagation length L (chip SOA input power $P(0) = P_{in}$). This leads to the definition of the single-pass chip gain G of the device,

$$P(L) = P(0)\exp\left(\int_0^L g\,dz\right) = P(0)\exp(gL) = GP(0), \qquad G = \frac{P(L)}{P(0)}. \qquad (2.18)$$

As mentioned before, also the real part n_{eff} of the effective refractive index is assumed to be independent of the propagation distance z, so that an integration of (2.17) yields the output phase $\phi(L)$ for a waveguide of length L (reference phase at SOA input $\phi(0) = 0$)

$$\phi(L) = \phi(0) + k_0 n_{\text{eff}} L, \qquad \phi(0) = 0. \qquad (2.19)$$

From (2.9), (2.18) and (2.19) it can be seen that the power increases exponentially with the SOA length, while the phase increases linearly only.

If the carrier density is increased from N_1 to N_2 due to carrier injection, the net modal gain increases by $\Delta g > 0$. The additional charge carriers slightly reduce the real part of the refractive index from n_{N1} to n_{N2}, so that $\Delta n < 0$. With the help of the alpha factor α_H (2.7), the change of the phase $\Delta\phi(L,N) = \phi(L,N_1) - \phi(L,N_2) =: \Delta\phi(N)$ at the output of the SOA can be expressed by the gain change

$$\Delta\phi(N) = k_0 \Gamma (n_{N1} - n_{N2})L = -k_0 \Gamma \Delta n L = -k_0 \Delta n_{\text{eff}} L = \alpha_H \,\Delta g L / 2. \qquad (2.20)$$

The absolute values of the gain must be derived from experiments by measurements of the input and output powers. Care must be taken to subtract the noise that may be superimposed. If noise cannot easily subtracted then a more complicated measurement approach may be used which is presented in *Section 2.2.3* "Noise Figure".

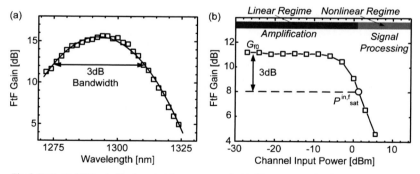

Fig. 2.11: Typical FtF gain G_{ff} characteristics as measured with fibers connected to the input and output of the SOA. (a) The bandwidth is around 40 nm with a peak gain of 15 dB at 1295 nm. The parameters have been derived at a temperature of 20°C, with a bias current of 490 mA and a continuous wave (cw) input power of −10 dBm. (b) Gain G_{ff} as a function of the input power. The in-fiber saturation input power level is indicated by a circle. The linear and the nonlinear operating regimes are shown, [B1].

Fig. 2.11 shows a typical dependency of the FtF gain G_{ff} on the (a) wavelength and (b) in-fiber input power $P^{\text{in,f}}$. Today, typical FtF gain values (expressed in dB as $10 \log_{10} G_{ff}$) between 10 dB and 30 dB are found in practice [21], [50]. The FtF gain G_{ff} is determined from the chip gain G by taking into account identical input and output fiber-to-chip coupling losses α_{Coupling} (per facet), thus, $G_{\text{ff}} = \alpha_{\text{Coupling}}\, G\, \alpha_{\text{Coupling}}$. In wavelength-division-multiplexing (WDM) systems signals on different wavelengths are simultaneously amplified which requires a large full width half maximum (FWHM) bandwidth of the SOA. From Fig. 2.11(a) it can be seen that this specific amplifier offers a 3 dB bandwidth B_G of 35 nm...40 nm. Gain bandwidths exceeding 100 nm are reported [24], [43], though.

In Fig. 2.11(b) also the 3 dB in-fiber saturation input power $P_{\text{sat}}^{\text{in,f}}$, which determines the input power at which the unsaturated FtF gain G_{f0} is halved, is indicated by a circle. The saturation input power is used to separate the regimes of so-called linear SOA and nonlinear SOA. A discussion on the saturation power follows in the respective *Section 2.2.4* on "Gain Saturation".

2.2.3 Noise Figure

SOA performance is also limited by spontaneous emission. This spontaneous emission adds to any input signal, it limits the overall number of amplifiers in a system and confines operation to signals with sufficient input powers.

If spontaneously emitted photons happen to propagate along the active region of an SOA, they are amplified and this noise is called amplified spontaneous emission (ASE) noise. The noise characteristic of an amplifier is generally described by a parameter called the noise factor (F) or in logarithmic units by the noise figure (NF), respectively [89]-[91].

In this paragraph, we will follow a noise factor definition which is valid for phase-insensitive amplifiers [92] that are discussed here. In a semi-classical interpretation for the case of reasonable gain values, we will obtain the noise factor/figure definition widely accepted within the optical amplifier community. This definition can also be used for amplifier cascades [91].

The noise figure is defined as the ratio of the (extractable, i.e., measurable) optical output noise power inside the optical bandwidth B_O and the (non-extractable, i.e., not directly measurable) optical power fluctuations of an ideal coherent signal in the same bandwidth. Equivalently, this ratio can be expressed by relating the signal-to-noise ratio (SNR) at the amplifier input to the SNR at the amplifier output,

$$F = \frac{\text{SNR}_{\text{in}}}{\text{SNR}_{\text{out}}}. \tag{2.21}$$

For the noise factor F, it is customary to use logarithmic units

$$\text{NF} = 10\log_{10}(F). \tag{2.22}$$

The average photon number corresponds to the average optical signal power P, which is proportional to the average photocurrent $\langle i \rangle$. The electrical power P_e is proportional to the square of the average photocurrent $\langle i \rangle^2$.

In the following, we assume ideally coherent signal light, the photon statistic of which is Poissonian. The associated quantum (shot) noise can be described by the variance $\sigma_i^2 = <\Delta i^2>_{\text{shot}}$ of the photocurrent i, where the photocurrent fluctuations $\Delta i = i - \langle i \rangle$ are defined with respect to the average current $\langle i \rangle$ [89].

We introduce the relation between the average photocurrent and the average optical signal power $\langle i \rangle = RP$, and define the sensitivity (responsivity) $R = \eta e / (hf_s)$ of a photodiode having a quantum efficiency η. Then, we write for the effective electrical photocurrent noise power in an electrical bandwidth B [90]

$$\left\langle \Delta i^2 \right\rangle_{\text{shot}} = 2e\langle i \rangle B = 2eRPB . \tag{2.23}$$

For a quantum (shot) noise limited reception the SNR at the input of the SOA is defined for a quantum efficiency $\eta = 1$ by

$$\text{SNR}_{\text{in}} = \frac{\langle i \rangle^2}{\left\langle \Delta i^2 \right\rangle_{\text{shot}}} = \frac{R^2 P_{\text{in}}^{\ 2}}{2eRP_{\text{in}}B} = \frac{P_{\text{in}}}{2hf_s B} . \tag{2.24}$$

However, the equivalent optical input noise power $2hf_s B$ cannot be measured directly. Two measurement techniques lead to the same result: Either SNR_{in} is determined by measuring P and by inferring the equivalent input noise power $2hf_s B$ from quantum theory, or the square $\langle i \rangle^2$ of the average photocurrent is measured along with its shot noise fluctuation $<\Delta i^2>_{\text{shot}}$, corrected for the real quantum efficiency η of the photodetector.

The output SNR from the SOA is given by the optical signal power at the amplifier output divided by the total optical noise power measured at the output in a bandwidth B. These quantities can be measured directly with an optical spectrum analyzer (OSA), or an electrical method can be employed, again using a photodetector and analyzing the resulting photocurrent.

The photocurrent fluctuations have to be subdivided into several contributions due to the fact that a photodiode is a square-law detector inducing mixing of the received components. Here, we consider contributions from the signal with shot noise (shot), as well as the beat-noise terms arising from signal-spontaneous (s-sp) noise and spontaneous-spontaneous (sp-sp) noise mixing [92].

To calculate the output SNR after the SOA, we assume that an optical filter is used prior to detection having an optical bandwidth B_O and a rectangular transfer function. This assumption is valid as the optical filters used in practice have bandwidths much narrower than typical SOA spectral bandwidths.

Furthermore, the ASE noise at the amplifier output is assumed to have a uniform optical power spectral density ρ_{ASE} over the filter bandwidth. The ASE noise power spectral density in one polarization only (here: $\rho_{\text{ASE}\parallel}$ copolarized with the signal) is $\rho_{\text{ASE}}/2$ and can be written with the inversion factor n_{sp} and $N_{1,2}$ as the number of excited microsystems in the lower and upper energy states, respectively [91],

$$\rho_{\text{ASE}} = 2\rho_{\text{ASE}\parallel} = n_{\text{sp}} 2hf_s(G-1) = 2P_{\text{ASE}\parallel} / B_O , \qquad n_{\text{sp}} = \frac{N_2}{N_2 - N_1} . \tag{2.25}$$

The different contributions to the photocurrent noise power in one polarization at the amplifier output are ($B_O \gg B$):

$$\left\langle \Delta i'^2 \right\rangle_{\text{shot}} = 2eRGP_{\text{in}}B , \tag{2.26}$$

$$\left\langle \Delta i'^2 \right\rangle_{\text{s-sp}} = 4R^2 G P_{\text{in}} \rho_{\text{ASE}\|} B, \tag{2.27}$$

$$\left\langle \Delta i'^2 \right\rangle_{\text{sp-sp}} = 2R^2 \rho_{\text{ASE}\|}^2 \left(B_{\text{O}} - B/2 \right) B \approx 2R^2 \rho_{\text{ASE}\|}^2 B_{\text{O}} B. \tag{2.28}$$

The square of the amplified signal current is $\left\langle i' \right\rangle^2 = R^2 G^2 P_{\text{in}}^2$. Then the output SNR is written

$$\text{SNR}_{\text{out}} = \frac{R^2 G^2 P_{\text{in}}^2}{2eRGP_{\text{in}}B + 4R^2 G P_{\text{in}} \rho_{\text{ASE}\|} B + 2R^2 \rho_{\text{ASE}\|}^2 B_{\text{O}} B}. \tag{2.29}$$

Thus, the noise factor F is obtained,

$$F = \frac{1}{G} + \frac{2\rho_{\text{ASE}\|}}{hf_s G} + \frac{\rho_{\text{ASE}\|}^2 B_{\text{O}}}{hf_s G^2 P_{\text{in}}} = \frac{1}{G} + \frac{2\rho_{\text{ASE}\|}}{hf_s G} + \frac{n_{\text{sp}} hf_s (G-1)}{hf_s G} \frac{P_{\text{ASE}\|}}{GP_{\text{in}}}. \tag{2.30}$$

This expression for the noise factor depends on the input signal. However, because n_{sp} is in the order of 1 and the signal output power GP_{in} after the SOA is typically much larger than the copolarized ASE power $P_{\text{ASE}\|}$, the last term in (2.30) can be safely neglected to yield the usual noise factor definition

$$F = F_{\text{shot}} + F_{\text{excess}} = \frac{1}{G} + \frac{2\rho_{\text{ASE}\|}}{hf_s G} = \frac{1}{G} + \frac{2P_{\text{ASE}\|}}{hf_s GB_{\text{O}}} = \frac{1}{G} + 2n_{\text{sp}} \frac{G-1}{G}, \tag{2.31}$$

which is independent of the input signal, obeys the cascading rules for amplifiers, and may be also derived from quantum mechanical first principles [90]. The FtF noise factor F_{ff} is obtained from the chip noise factor F by taking into account the input coupling losses $F_{\text{ff}} = \alpha_{\text{Coupling}} F$.

From (2.31) it can be seen that the minimum achievable excess noise figure of an optical amplifier is $F = 2$ (NF = 3 dB) for the case of maximum inversion $n_{\text{sp}} = 1$ and a large single-pass gain ($G \gg 1$).

In specific network applications it is of interest to cascade SOA in a link. To evaluate the total noise factor F_{tot} of the amplifiers in the link a number of m_{OA} SOA is considered, each with gain $G_q \geq 1$ and excess noise factor F_q ($q = 1 \dots m_{\text{OA}}$), followed by a fiber section with a linear "gain" of $0 < a_q \leq 1$. Assuming narrowband optical filters after each SOA, F_{tot} can be written as

$$F_{\text{tot}} = F_{\text{shot}}^{\text{total}} + F_{\text{excess}}^{\text{total}} = \frac{1}{\prod\limits_{q=1}^{m_{\text{OA}}} G_q a_q} + \left[F_1 + \frac{F_2}{G_1 a_1} + \dots + \frac{F_{m_{\text{OA}}}}{\prod\limits_{q=1}^{m_{\text{OA}}-1} G_q a_q} \right] \tag{2.32}$$

Assuming $a_q = 1$ and $m_{\text{OA}} = 2$ in (2.32), it can be seen that a low-noise, high output amplifier can be constructed by combining a low-noise high-gain first stage amplifier section followed by a high output power amplifier, the noise of which can be larger. If a cascade of identical SOA is investigated in which the SOA gain in each span compensates for the losses in each span ($G_q a_q = 1$), the total noise factor results to $F_{\text{tot}} = 1 + m_{\text{OA}} F$.

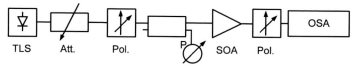

Fig. 2.12: Setup to measure the gain and the noise figure of an SOA as a function of different input power levels and wavelengths. The wavelength-tunable laser source (TLS) provides a continuous-wave signal. The input power to the SOA is varied by an optical attenuator (Att.) and measured with a power meter (P) after a polarizer (Pol.) and a 50:50 coupler. The output signal power and the amplified spontaneous emission (ASE) power are measured by an optical spectrum analyzer (OSA). The second polarizer transmits only ASE noise which is co-polarized with the signal, [B1].

Of important practical interest is the measurement technique to determine the gain and the noise figure of an SOA. The measurement setup is depicted in Fig. 2.12. A cw signal from a tunable laser source (TLS) is fed to an attenuator (Att.) which allows the accurate control of the SOA input power, measured with a power meter (P) and a 50:50 coupler. The input polarization to the SOA is adjusted with a polarizer (Pol.). After the SOA, the output signal power and the ASE power are measured with an optical spectrum analyzer (OSA). The second polarizer enables the measurement of the ASE noise which is co-polarized with the signal only.

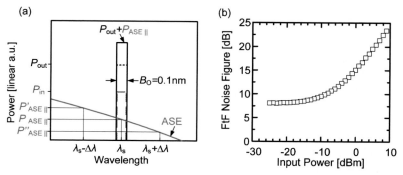

Fig. 2.13: Schematic for determining gain $G = P_{out} / P_{in}$, ASE noise power and noise figure (NF) of an SOA with typical NF_{ff} measurement results. Spectral data refer to a resolution bandwidth $B_O = 0.1$ nm. (a) Measured input power P_{in}, total output power $P_{out} + P_{ASE\|}$, and spectral dependance of ASE noise power. The actual coherent output power P_{out} is retrieved by subtracting the ASE noise power $P_{ASE\|}$, which is obtained by a linear interpolation of the ASE noise powers $P'_{ASE\|}$ and $P''_{ASE\|}$ measured in a distance $\Delta\lambda$ from the coherent carrier wavelength λ_s (here the SOA chip values are used to simplify the figure). (b) Measured NF_{ff} increases for increasing input power due to a decreasing gain, [B1].

In Fig. 2.13(a) the measurement principle is explained. The OSA detects the output power $P_{out} + P_{ASE\|}$ at the coherent-carrier wavelength λ_s in a resolution bandwidth B_O usually set to an equivalent wavelength interval of 0.1 nm. To separate P_{out} and P_{ASE}, the co-polarized ASE powers $P'_{ASE\|}$ and $P''_{ASE\|}$ are measured in a distance of $\Delta\lambda = 1$ nm to the left and the right of λ_s. A linear interpolation $P'_{ASE\|} + P''_{ASE\|} / 2$ gives an estimate of the ASE power $P_{ASE\|}$ at λ_s.

From the measured quantities P_{in} and P_{out}, the gain G can be calculated in addition, (2.18). From G, $P_{ASE\parallel}$ and B_O the noise figure (2.31) is calculated.

Fig. 2.13(b) shows a typical result of a FtF noise figure characterization. A noise figure of $NF_{ff} = 7$ dB for low in-fiber input powers $P^{in,f}$ is found. The noise figure increases for increasing input powers due to the associated decrease in single-pass gain. Commercially, SOA are available with FtF noise figures in the region of $NF_{ff} = 5...10$ dB [93], [94].

2.2.4 Gain Saturation

Since an SOA provides a limited amount of extractable power at a constant bias current only, the gain reduces if the SOA input power is sufficiently large. To describe this so-called gain saturation, the saturation power of an SOA is introduced in this section.

With the help of a rate-equation model that governs the interaction of photons and electrons inside the active region, the so-called saturation input power is calculated. The saturation input power of an SOA describes the input power at which the unsaturated single-pass chip gain reduces by 3 dB. In this section, the rate equations are specified heuristically by considering the phenomena through which the number of carriers changes with time inside the active volume. The dominant process is the electronic carrier injection and the resultant effective carrier lifetime τ_c, i.e., the time for an inter-band transition. Under steady-state operation, the injection and recombination of carriers are in equilibrium and occur on a nanosecond to hundreds of picoseconds timescale. Intra-band phenomena which can also contribute to gain saturation are disregarded here, but are considered in the following section on "SOA dynamics". The model can be considerably simplified by assuming an SOA input signal with a pulse width τ_p much longer than the effective carrier lifetime $\tau_c \approx 1$ ns. Then the intra-band dynamics of carriers (leading to an effective intra-band relaxation time τ_{intra}) can be neglected since the associated intra-band relaxation timescale is typically in the range of hundreds of femtoseconds up to a few picoseconds. The assumption $\tau_p \gg \tau_c \gg \tau_{intra}$, is typically justified for steady-state (continuous wave) or low bit rate (large pulse width) operation. Furthermore, longitudinal and transverse carrier diffusion is neglected, and a uniform distribution of the bias current I across the active region is assumed [95].

Equation (2.33) relates the temporal change of the carrier density N to the carrier injection rate I/e per active volume $V = CL = wdL$ (waveguide cross-section area C, active region length L) and to the carrier recombination rates R_c and R'_s, which are due to spontaneous recombination (radiative and non-radiative [96]), and to stimulated (radiative) transitions, respectively. The stimulated recombination rate due to the amplification of the signal is R_s, and the recombination rate due to ASE noise is R_{ASE}. The quantity τ_c is the so-called effective carrier lifetime, v_g the group velocity, g the net modal gain and S' the total photon density. The total photon density S' consists of the signal photon density S and the ASE photon density S_{ASE}. Because of charge neutrality the carrier density N represents both, the electron and the hole concentration. Further, it needs to be mentioned that $v_g g = G'$ is the gain rate. The carrier rate equation reads

$$\frac{dN}{dt} = \frac{I}{eV} - R_c - R_s' = \frac{I}{eV} - \frac{N}{\tau_c} - g v_g S', \qquad R_s' = R_s + R_{ASE}, \qquad S' = S + S_{ASE}. \qquad (2.33)$$

We assume that the net modal gain g [unit 1/cm] depends linearly on the density of carriers injected from an external current supply. The quantity a is the differential gain, and N_t is the carrier density at transparency. The net modal gain can be approximated by

$$g \approx \frac{dg}{dN}(N - N_t) = a(N - N_t), \qquad a = \frac{dg}{dN}. \qquad (2.34)$$

This linear relation is a good approximation at a wavelength corresponding to the gain peak of the SOA [95]. If the SOA is operated at other spectral positions inside its gain bandwidth, (2.34) can be replaced by a polynomial model with quadratic and cubic dependencies on the carrier density and wavelength [32], [82].

Saturation Input Power. For small input signal powers, the output power of an amplifier depends linearly on the input power, if we disregard noise as is the case for the following. Beyond the so-called saturation input power, the output power increases sub-linearly (it "saturates"), because the gain starts decreasing ("gain suppression"). Gain suppression is explained by the fact that the carrier injection rate which is fixed by the external current source sets a limit to the number of electron-hole pairs that can possibly recombine per time. Therefore, the rate of generated photons is also limited.

Using (2.34) and rewriting the total photon density S averaged over the active volume in terms of an average optical power $\Gamma P = S \, hf_s \, v_g C$, the relation (2.33) can be written as

$$\frac{dN}{dt} = \frac{I}{eV} - \frac{N}{\tau_c} - g v_g S = \frac{I}{eV} - \frac{N}{\tau_c} - \Gamma a (N - N_t) \frac{P}{C hf_s}. \qquad (2.35)$$

Equation (2.35) is solved for a time-independent carrier concentration $dN/dt = 0$ by

$$N = \frac{\tau_c I}{eV} \frac{P_s}{P_s + P} + N_t \frac{P}{P_s + P}, \qquad P_s = \frac{C hf_s}{a \Gamma \tau_c}. \qquad (2.36)$$

Substituting (2.36) in (2.34) modifies (2.16) to

$$\frac{dP}{dz} = a \left(\frac{\tau_c I}{eV} - N_t \right) \frac{P_s}{P_s + P} P = g(P) P, \qquad g(P) = a \left(\frac{\tau_c I}{eV} - N_t \right) \frac{P_s}{P_s + P},$$

$$\frac{dP}{dz} = g_0 \frac{P_s}{P_s + P} P, \qquad g = g_0 \frac{1}{1 + \dfrac{P}{P_s}}, \qquad g_0 = a(N_0 - N_t), \qquad N_0 = \frac{\tau_c I}{eV}, \qquad (2.37)$$

$$g = g_0 \frac{1}{1 + \dfrac{S}{S_s}} = \frac{g_0}{1 + \varepsilon_c S}, \qquad \varepsilon_c = \frac{1}{S_s} = v_g a \tau_c.$$

For small input powers $P \ll P_s$, the quantities g_0 and N_0 represent the small-signal net modal gain and the average carrier concentration in analogy to the more general relation (2.34). From (2.37), it can be seen that the parameter P_s is the power level at which the power-dependent net modal gain $g = g(N) = g(N(S))$ is half of its unsaturated value g_0. As an alternative to the saturation power P_s, the nonlinear gain compression factor ε_c (the inverse saturation photon density S_s) can be used which transforms (2.35) then into

$$\frac{dN}{dt} = \frac{I}{eV} - \frac{N}{\tau_c} - v_g gS = \frac{I}{eV} - \frac{N}{\tau_c} - v_g \frac{g_0 S}{1+\varepsilon_c S}. \tag{2.38}$$

Equation (2.37) can be solved, employing the boundary conditions $P(z=0)=P_{in}$ and $P(z=L)=P_{out}$, and substituting the single-pass chip gain G from (2.18) and the unsaturated single-pass chip gain G_0,

$$\frac{P_{out}}{P_{in}} = G = G_0 \exp\left[-(G-1)\frac{P_{in}}{P_s}\right], \qquad G_0 = \exp(g_0 L). \tag{2.39}$$

From (2.37), we see that as long as the SOA output power GP_{in} with $G \geq 1$ is much lower than the saturation power, $GP_{in} \ll P_s$, the first exponential term in (2.39) is close to 1 and the amplifier gain is almost G_0. However, when the output power exceeds the saturation power, the amplifier gain decreases. The input P_{sat}^{in} and output $P_{sat}^{out} = (G_0 / 2)P_{sat}^{in}$ chip saturation powers, for which the unsaturated single-pass chip gain is halved, can be calculated to be

$$P_{sat}^{in} = \frac{2\ln 2}{G_0 - 2} P_s = \frac{2hf_s \ln 2}{G_0 - 2} \frac{C}{\Gamma} \frac{1}{a} \frac{1}{\tau_c}, \tag{2.40}$$

$$P_{sat}^{out} = \frac{G_0 \ln 2}{G_0 - 2} P_s = \frac{hf_s \ln 2 G_0}{G_0 - 2} \frac{C}{\Gamma} \frac{1}{a} \frac{1}{\tau_c}. \tag{2.41}$$

The saturation input or output powers can be easily determined from a single-pass gain measurement. Typical values for the in-fiber saturation input power $P_{sat}^{in,f}$ are -15 dBm to 0 dBm depending on the unsaturated single-pass FtF gain and the saturation power of the device.

Unlike linear SOA, nonlinear SOA are optimized for very low input saturation powers. This helps performing all-optical switching at highest-speed with low input powers. Using ideal nonlinear configurations record low optical switching powers of -17.5 dBm or 0.45 fJ per bit at 40 Gbit/s has already been shown [71].

2.2.5 SOA Dynamics

For a linear amplifier the gain does not depend on the amplifier input power, if it is much smaller than the saturation input power P_{sat}^{in}. Therefore, in an arbitrary sequence of "1"s and "0"s, each bit experiences the same small-signal gain, and there is no dependency on the bit pattern. Similarly, if multiple signals with differing wavelengths are amplified simultaneously, there is no cross-talk between the wavelength channels. In both cases, the situation would be different if larger average input powers would lead to carrier depletion and therefore to a gain lower than the small-signal gain. A series of "1"s would lead to a temporal gain decrease, and high average input power in one wavelength channel would reduce the gain for signals at a different wavelengths. This may result in undesirable signal patterning and wavelength channel cross-talk. It should be noticed though, that such cross-talk may be exploited for nonlinear optical operations such as 2R [66] and 3R all-optical wavelength conversion [69], [97], for performing logical optical operations [98] or all-optical demultiplexing [99].

Even if the input power exceeds the saturation input power $P_{\text{sat}}^{\text{in}}$, a quasi-linear operation is possible, depending on the gain dynamics and therefore depending on the temporal change of the charge carriers. If the SOA carrier recovery between subsequent "1"s is sufficiently fast, then the gain for each "1"-bit remains same. On the other hand, if the amplifier dynamics are sufficiently slow, an average input power larger than $P_{\text{sat}}^{\text{in}}$, leads to a stationary gain compression. If "1"s and "0"s are equally distributed, the gain would remain constant, but at a level smaller than the small-signal gain. This is the case in, e.g., doped fiber amplifiers which show fluorescence lifetimes of a few milliseconds [100]. Thus, such fiber amplifiers are successfully used in today's fiber communications systems in which the transmission bit-period of, e.g., a 10 Gbit/s data signal is 100 ps.

In the case of an SOA the situation is more involved since the typical gain recovery time is in the order of 100 ps. Therefore, if SOA are used to amplify high bit rate (> 10 Gbit/s) data signals, bit-pattern effects and channel cross-talk will occur if the input power exceeds $P_{\text{sat}}^{\text{in}}$. In the following section, the SOA gain recovery due to slow inter-band and fast intra-band processes is discussed.

Inter-band Effects

Typical effective carrier lifetimes for spontaneous inter-band transitions are in the range of $\tau_c = 100$ ps... 1 ns. Assume a pulse sequence where the bit-period is of the order of the impulse width τ_p which is much larger than the effective carrier lifetime, $\tau_p \gg \tau_c$. If the power of an input pulse is sufficiently large so that it reduces the SOA gain, the time to re-establish the gain to its original level is of interest. This time is called gain recovery time and will be specified in the following.

The SOA carrier density evolution is described by the rate equation (2.35) for the case that only inter-band effects are considered. Since the carrier density N and the net modal gain g are assumed to be linearly related as in (2.34), the gain evolution can be derived from (2.35), too. To simplify the description, we combine the effective lifetime τ_c for radiative and non-radiative spontaneous recombinations with the lifetime $\tau_{s'}$ for stimulated recombination to a *total* effective carrier lifetime $\tilde{\tau}_c$ [101],

$$\frac{1}{\tilde{\tau}_c} = \frac{R_c + R_s'}{N} = \frac{R_c}{N} + \frac{R_s + R_{\text{ASE}}}{N} = \frac{1}{\tau_c} + \frac{1}{\tau_{s'}} \ , \qquad R_s \sim S, \ R_{\text{ASE}} \sim S_{\text{ASE}}. \tag{2.42}$$

The value of $\tilde{\tau}_c$ strongly depends on the SOA operating point, i.e., on the carrier concentration N.

The effective carrier lifetime τ_c is determined by non-radiative recombination via impurity levels in the forbidden band (Shockley-Read-Hall recombination, subscript SRH), by inter-band radiative spontaneous recombination (subscript sp), by non-radiative Auger recombination (subscript Au), and by a term that takes carrier leakage through the active region into account (subscript Leak) [82]. The terms have the corresponding coefficients A_{SRH}, B_{sp}, C_{Au} and D

$$\frac{1}{\tau_c} = \frac{1}{\tau_{SRH}} + \frac{1}{\tau_{sp}} + \frac{1}{\tau_{Au}} + \frac{1}{\tau_{Leak}} = A_{SRH} + B_{sp}N + C_{Au}N^2 + DN^{5.5}. \tag{2.43}$$

The dominant contribution in (2.42) depends strongly on the SOA operating point. For short SOA lengths (no significant ASE) and zero input power (no amplified signal), the total effective carrier lifetime $\tilde{\tau}_c$ is dominated by the effective carrier lifetime $\tau_c = N/R_c$ due to spontaneous radiative and non-radiative recombinations. For long SOA length (significant ASE) and zero input power, the ASE recombination rate $R_{ASE} \sim S_{ASE}$ dominates. If an input signal is present, the stimulated emission rate due to the signal $R_s \sim S$ dominates.

From a practical point of view, the total effective carrier lifetime $\tilde{\tau}_c$ is not used to determine the SOA speed limitation in system experiments. This is because the SOA gain recovery time should also take into account the strength of gain saturation. After an abrupt transition from the small-signal chip gain G_0 to a suppressed gain G_{supp}, a 90 %-10 % gain recovery time $\tau_{90\%-10\%}$ measures the effective rise time of the gain towards G_0. It is defined by the difference of the time at which the actual gain is $G_{90\%} = G_0 - 0.9 \times (G_0 - G_{supp})$ and the time where the gain is re-established to $G_{10\%} = G_0 - 0.1 \times (G_0 - G_{supp})$.

The gain recovery time $\tau_{90\%-10\%} = |t(G_{90\%}) - t(G_{10\%})|$ (see Fig. 2.14) is related to the total effective carrier lifetime. In detail, one finds [101]

$$\tau_{90\%-10\%} \approx \tilde{\tau}_c \ln\left[\frac{\ln(0.1 + (0.9G_{supp}/G_0))}{\ln(0.9 + (0.1G_{supp}/G_0))}\right],$$

$$\lim_{G_0 \gg 9G_{supp}} \tau_{90\%-10\%} = \tilde{\tau}_c \ln\left(\frac{\ln 0.1}{\ln 0.9}\right) \approx 3\tilde{\tau}_c. \tag{2.44}$$

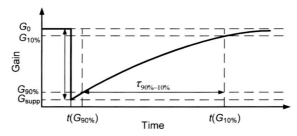

Fig. 2.14: Schematic for determining the 90%-10% gain recovery time. After an abrupt transition from the small-signal gain G_0 to a suppressed gain G_{supp}, a 90 %-10 % gain recovery time $\tau_{90\%-10\%} = |t(G_{90\%}) - t(G_{10\%})|$ measures the effective rise time of the gain back towards G_0, [B1].

From (2.44) it can be seen that the gain recovery time depends on the degree of gain suppression. A short total effective carrier lifetime implies faster gain recovery, while a larger gain suppression G_0/G_{supp} leads to a slower recovery. For $G_0/G_{supp} \gg 9$, $\tau_{90\%-10\%}$ approaches $3\tilde{\tau}_c$.

Ordinarily, the gain recovery time is in the range of a few picoseconds (ps) up to several hundreds of ps. It has been found that a couple of mechanisms can be used to reduce this gain recovery time for an SOA. For example, a bias current increase reduces the recovery time, because the increase in the carrier number leads to a larger R_c [102]. For a constant net modal gain g, an increase in the device length increases the gain $G = \exp(gL)$. Therefore the spectral density of the ASE noise (2.25) increases, and the resulting larger recombination rate $R_{ASE} \sim S_{ASE} \sim \rho_{ASE}$ (2.42) reduces the recovery time [103], [104]. Another technique to reduce the recovery time is the use of a holding beam to enhance the stimulated emission rate R_s [102], [105]. Further, the SOA material properties are often tailored by n-doping or p-doping of the active region to increase R_c [106].

In recent years, it was found that the use of QD in the active SOA region can also reduce the recovery time down to a few ps. The QD are embedded in a so-called wetting layer which acts as a carrier reservoir for the QD states [23], [J10]. In principle, by a proper population of the reservoir states, the fast refilling of QD states from the wetting layer dominates the gain recovery process. The refilling of wetting layer states follows the physics as discussed above and is ultimately limited by the supply rate of carriers through the injection current.

Intra-band Effects

Fast intra-band transitions come into play if pulses with a duration shorter than the effective carrier lifetime τ_c are investigated. Intra-band effects are characterized by a non-equilibrium carrier distribution within the bands, whereas inter-band effects are characterized by a non-equilibrium carrier distribution between the conduction and the valence band.

Before an optical pulse is launched to a forward biased SOA, the CB and VB carriers inside each band are in equilibrium, Fig. 2.15(b), and follow a Fermi-Dirac distribution. When an optical beam with a short duration (few ps) is launched in the SOA, CB electrons at the appropriate photon energies are depleted because of stimulated recombinations [107]-[109]. Therefore, at these photon energies, the carrier number and consequently the gain is reduced, Fig. 2.15(a) and Fig. 2.15(b). A (nearly symmetrical) spectral "hole" is "burned" into the carrier distribution, and therefore this effect is called spectral hole burning (SHB). It is an ultra-fast effect on a time scale of a few femtoseconds (fs).

In addition, pairs of photons are absorbed, a process which generates an electron-hole pair. For this so-called two-photon absorption (TPA), the sum of the two photon energies is larger than the bandgap energy. The TPA-generated free carriers absorb light (free-carrier absorption, FCA) [101] and move to higher energy states within the same band. SHB, TPA and FCA alter the carrier distributions from intra-band equilibrium to non-equilibrium, so that the Fermi-Dirac distribution function is not applicable any more. SHB eliminates carriers which need to be refilled by carrier-carrier scattering from higher energy states, so that the carrier distributions flattens out and the effective carrier temperature rises. Furthermore, carriers hotter than the average are added by TPA and FCA, thereby additionally increasing the average carrier energy beyond the one associated with the lattice temperature [82]. The process is named carrier heating (CH in Fig. 2.15(a)) and occurs on a time scale of 100 fs (i.e., 78 fs in a 1.55 μm bulk SOA or 95 fs in a 1.3 μm bulk SOA [33]]) named the SHB relaxation time τ_{SHB}. The hot carriers cool down (carrier cooling (CC) in Fig. 2.15(a)) by carrier-phonon scattering on a time scale of up to several ps characterized by a CH relaxation time τ_{CH} (i.e., 700 fs in a 1.55 μm bulk SOA or 1000 fs in a 1.55 μm MQW SOA [33]).

Finally, carrier injection fills the depleted states on a nanosecond scale (CI in Fig. 2.15(a)), until the initial intra-band equilibrium state is reached eventually. The associated time constant is the effective carrier lifetime τ_c.

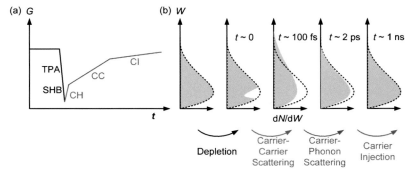

Fig. 2.15: Gain as a function of time and explanation of intra-band effects, modified from [107]. (a) Gain suppression due to spectral hole burning and two-photon absorption. Gain recovery is caused by carrier heating, carrier cooling and carrier injection. (b) The corresponding energy versus the density of electrons N per energy interval in the conduction band, [B1].

In the following, we incorporate short-time intra-band carrier effects into a theory of the SOA carrier dynamics. We distinguish two cases related to the pulse width τ_p. First, if the pulse width is smaller than the effective carrier lifetime τ_c and larger than the effective intra-band relaxation time τ_{intra}, then on the scale of the pulse width the intra-band effects occur instantaneously. We include the various intra-band effects phenomenologically by a gain compression that becomes the more pronounced the larger the photon density S is, see (2.37). Second, if the pulse width is comparable to the effective intra-band relaxation time, intra-band memory effects must be taken into account by rate equations.

Case 1, $\tau_c \gg \tau_p \gg \tau_{intra}$. Here, we consider saturation by SHB and CH with its associated relaxation times [108], [110], [111]. For this case, the nonlinear gain compression factor ε_c for inter-band effects in (2.37) is supplemented with the corresponding factors ε_{SHB} and ε_{CH} for SHB and CH [108], [110], [111]

$$g = \frac{g_0}{1 + (\varepsilon_c + \varepsilon_{SHB} + \varepsilon_{CH})S}. \tag{2.45}$$

The rate equation (2.35) for the time dependance of the carrier density N is then modified by the power-dependent net gain $g(S)$ from (2.45) in analogy to (2.38),

$$\frac{dN}{dt} = \frac{I}{eV} - \frac{N}{\tau_c} - v_g g S = \frac{I}{eV} - \frac{N}{\tau_c} - v_g \frac{g_0 S}{1 + (\varepsilon_c + \varepsilon_{SHB} + \varepsilon_{CH})S}. \tag{2.46}$$

This approach is often used to model the behavior of the SOA dynamic in system experiments for up to 100 Gbit/s on-off keying data signals.

Case 2, $\tau_\text{p} \sim \tau_\text{intra}$. Now the data pulse width of interest is in the order of the effective intra-band relaxation times [110]. Intra-band effects modify the net modal gain from $g(N(t)) = g(t)$, see (2.34) and (2.37), to $g_\text{tot}(t)$, so that the time-dependent gain now reads

$$G(t) = \exp\left(\int_0^L g_\text{tot}(t)\,\mathrm{d}z \right) = \exp\left(g_\text{tot}(t)L \right), \qquad (2.47)$$

and (2.35), (2.38) changes to

$$\frac{\mathrm{d}N}{\mathrm{d}t} = \frac{I}{eV} - \frac{N}{\tau_\text{c}} - v_\text{g} g_\text{tot} S . \qquad (2.48)$$

It is convenient to introduce a dimensionless gain coefficient $h_\text{tot}(t)$, which is expressed as a sum of the bias-dependent inter-band term $h(t) = g(t)L$ and the additional contributions of SHB and CH to the compression of the gain, $\Delta h_\text{SHB}(t) = \Delta g_\text{SHB}(t)L$ and $\Delta h_\text{CH}(t) = \Delta g_\text{CH}(t)L$, respectively. Following [101], [110] we write for $h_\text{tot}(t)$

$$h_\text{tot}(t) = \int_0^L g_\text{tot}(t)\,\mathrm{d}z = g_\text{tot}(t)L = h(t) + \Delta h_\text{SHB}(t) + \Delta h_\text{CH}(t), \quad G(t) = \exp\left(h_\text{tot}(t) \right). \quad (2.49)$$

The time-dependance of the integrated modal gain $h(t)$ for the inter-band effects is determined as before by the dynamics of the carrier density N which is given by (2.35). The corresponding rate equation for $h(t)$ is derived by starting from the time dependent rate equation for the carrier density N (2.35). We substitute N by the net modal gain g from (2.34) and integrate the resulting differential equation over the SOA length L. We then re-write the differential equation for the spatially dependent power (2.16) in terms of the photon density S, where the total gain g_tot replaces g which was determined by inter-band effects only. After an integration over the SOA length we end up with a differential equation for $h(t)$. Having introduced the SOA input photon density $S = S_\text{in}(t)$ and the nonlinear gain compression factor $\varepsilon_\text{c} = v_\text{g} a \tau_\text{c}$ from (2.37), we find

$$\tau_\text{c} \frac{\mathrm{d}h(t)}{\mathrm{d}t} = h_0 - h(t) - \left(G(t) - 1 \right) \varepsilon_\text{c} S_\text{in}(t), \qquad h_0 = g_0 L, \qquad \frac{\mathrm{d}S(z)}{\mathrm{d}z} = g_\text{tot} S(z),$$

$$S_\text{out}(t) = S(t,L) = G(t)S(t,0) = G(t)S_\text{in}(t), \qquad \int_0^L g_\text{tot} S\,\mathrm{d}z = \left(G(t) - 1 \right) S(t,0). \qquad (2.50)$$

The rate equation for $h(t)$ can be directly compared to (2.33) and (2.35). Via (2.34), the proportionality $h \sim g \sim N$ holds, and $h_0 \sim g_0 \sim N_0 \sim I$ is connected to the fixed bias current I. The induced emission term incorporates all gain compression effects in its total gain $G(t)$. If for a small integrated total modal gain $h_\text{tot} \ll 1$ the approximation $G(t) - 1 = \exp\left(h_\text{tot}(t) \right) - 1 \approx h_\text{tot}(t)$ holds true, the photon density $S(z)$ does not change significantly along z. Therefore, the integral over z needs not to be executed, and an approximation to (2.50) follows directly by substituting (2.33) in (2.48).

The intra-band effects are described by nonlinear gain compression factors ε_SHB and ε_CH for SHB and CH, respectively. Conduction band carriers are considered, and effects such as TPA and FCA are neglected. Using the SHB relaxation time τ_SHB and the CH relaxation time τ_CH we write, following the references [82], [101], [109], [110],

$$\tau_{\text{SHB}} \frac{\partial \Delta h_{\text{SHB}}}{\partial t} = -\Delta h_{\text{SHB}} - [G(t)-1]\varepsilon_{\text{SHB}} S_{\text{in}}(t) - \tau_{\text{SHB}} \left(\frac{\partial \Delta h_{\text{CH}}}{\partial t} + \frac{\partial h}{\partial t} \right), \qquad (2.51)$$

$$\tau_{\text{CH}} \frac{\partial \Delta h_{\text{CH}}}{\partial t} = -\Delta h_{\text{CH}} - [G(t)-1]\varepsilon_{\text{CH}} S_{\text{in}}(t). \qquad (2.52)$$

Carrier heating due to an input impulse S_{in}, which instantaneously burns a spectral hole, reduces the integrated modal gain by $|\Delta h_{\text{CH}}|$ according to (2.52); $|\Delta h_{\text{CH}}|$ then relaxes back to zero due to carrier cooling on a time scale τ_{CH}. On the other hand, the spectral hole burned by the input impulse reduces the integrated modal gain by $|\Delta h_{\text{SHB}}|$ according to (2.51); $|\Delta h_{\text{SHB}}|$ then relaxes back to zero due to carrier heating on a time scale τ_{SHB}. If carrier cooling has a small effect, this relaxation time is large (see last but one term in (2.51)). The same is true if carriers are injected at a low rate only, (see last term in (2.51)).

2.2.6 Alpha Factor

So far, the alpha factor α_{H} was assumed to be time-independent. This implies that at the timescale considered in (2.7), refractive index changes and gain changes were identically related to a carrier concentration change. Following a carrier depletion, the total number of carriers recovers on a time scale fixed by the external injection rate, and so does the refractive index n_{N}. However, as was discussed in the previous section, there are various time scales for gain recovery. As a consequence, the alpha factor becomes time dependent and is renamed to be an effective time-dependent alpha factor $\alpha_{\text{eff}}(t)$. This approach is contrary to most publications in which the alpha factor is tacitly assumed to be constant in time. For linear low data rate applications, i.e., for a 10 Gbit/s FTTH reach extender, this approximation is good enough. However, for describing SOA at highest data rates (> 100 Gbit/s) and smallest pulse widths, the time-dependent alpha-factor approach offers higher accuracy. More details on time dependency and modeling of the effective alpha factor for wavelength conversion can be found in reference [112]. Typical reported values for SOA are in the range of $\alpha_{\text{H}} = 0...2$ for QD SOA and $\alpha_{\text{H}} = 2...8$ for QW or bulk SOA.

In the following the time dependance of the SOA phase change $\Delta\phi(N(t))$ resulting from carrier and subsequent gain changes is discussed. Starting from the definition (2.7) and in the spirit of the previous section, we interpret the integrated net modal gain change ΔgL as the change $\Delta h_{\text{tot}}(t)$ of the time-dependent integrated modal gain $h_{\text{tot}}(t)$ in (2.49). In analogy to (2.20) we write

$$\Delta\phi(t) = \alpha_{\text{eff}}(t)\Delta g_{\text{tot}}(t)L/2 = \alpha_{\text{eff}}(t)\Delta h_{\text{tot}}(t)/2, \quad \Delta g_{\text{tot}}(t)L = \Delta h_{\text{tot}}(t). \qquad (2.53)$$

The time-dependent effective alpha factor is defined as

$$\alpha_{\text{eff}}(t) = \frac{2\Delta\phi(t)}{\Delta h_{\text{tot}}(t)} = \frac{2\Delta\phi(t)}{\Delta(\ln G(t))}. \qquad (2.54)$$

A detailed discussion of the constituents of $\alpha_{\text{eff}}(t)$ comprising SHB (α_{SHB}), carrier heating (α_{CH}) and band-filling from carrier injection (α_{H}), and the respective changes of the effective refractive indices Δn_{SHB}, Δn_{CH}, Δn_{eff} together with the changes in the gain Δg_{SHB}, Δg_{CH}, Δg is found in [112]. The contribution α_{SHB} is almost zero since SHB produces a nearly symmetrical spectral hole centered at the signal wavelength, so that the Kramers-Kronig integrand becomes anti-symmetric around the operating frequency, and the Kramers-Kronig integral remains small. The contribution α_{CH} may be approximated by a constant [112]. The band-filling term α_H finally shows a more complicated behavior and it strongly depends on the signal wavelength as well as on the carrier density N. An appropriate parameterization to model this alpha factor as a function of wavelength and carrier density may be found in [82]. In summary, it is to be expected that initially the alpha factor is small due to SHB, and that it increases for larger times.

Pump-Probe Measurement Setup and Results
For determining gain and phase recovery times of an SOA, a heterodyne pump-probe technique with sub-picosecond resolution has been employed as shown in Fig. 2.16 [J10].

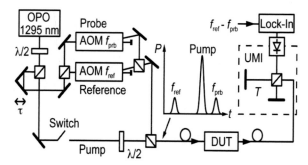

Fig. 2.16: Schematic of the heterodyne pump-probe setup. Short pulses are generated by an optical parametric oscillator (OPO) and split into pump, probe and reference pulses. Probe and reference pulses are tagged by a frequency shift f_{prb} and f_{ref}, respectively, which is induced by acousto-optic modulators (AOM). A strong pump pulse drives the SOA into its nonlinear regime. A weak pulse probes these nonlinearities in gain and phase. After the device under test (DUT), the pulse train is split and recombined in a Michelson interferometer with unbalanced arm lengths (UMI) such that the resulting probe-reference beat signal with frequency $f_{ref} - f_{prb}$ can be detected in amplitude and phase by a lock-in amplifier, [J10] and [B1].

Pulses with a FWHM of 130 fs are generated by an optical parametric oscillator (OPO) and split into pump, probe and reference pulses. Probe and reference pulses are tagged by frequency shifts f_{prb} and f_{ref}, respectively, which are induced by acousto-optic modulators (AOM). The pulse repetition rate is 80 MHz. The pulses are coupled into a polarization-maintaining (PM) fiber, from where they are launched into the SOA. First, the weak reference pulse is guided through the unperturbed device under test (DUT). Next, a strong pump pulse drives the SOA into its nonlinear, gain-depleted regime. A weak pulse probes these nonlinearities in gain and phase. After the device, the pulse train is split in a Michelson interferometer with unbalanced arm lengths. Both copies of the pulse train are then superimposed in such a way that the reference pulse leaving the long arm coincides with the probe pulse leaving the short arm (UMI). Because of the respective frequency shifts, the lock-in amplifier detects amplitude and phase of the photodiode current at the difference frequency $f_{ref} - f_{prb}$. By varying the optical delay between -10 ps and 300 ps, the temporal evolution of gain and phase can be measured.

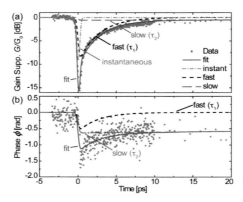

Fig. 2.17: Time evolution of (a) gain suppression G/G_0 and (b) phase dynamics of a typical QD SOA device. The fit (solid line) well reproduces the measured data (gray dots). The model assumes a fast process (dashed line), a slow process (long-dashed line) and instantaneous two-photon absorption (TPA, dash-dotted line) [J10] and [B1].

The typical gain and phase dynamics of an SOA (here especially a QD SOA) are shown in Fig. 2.17. However, in pump-probe experiments, the number of measurable time constants is limited by temporal resolution and by noise. Therefore, it is only possible to draw conclusions regarding the dominant relaxation processes, the time dependance of which is usually approximated by an exponential function. Here, two processes with different relaxation times are found: A fast one associated with the quantum-dot carrier capture time τ_1 in the order of a few ps (comparable to τ_{SHB}), and a slow one associated with the wetting layer capture time τ_2 in the order of several hundred ps (comparable to τ_c). Both processes influence the material gain and the phase change simultaneously, but with different magnitudes.

As already discussed in Fig. 2.14, the 90 %-10 % recovery time $\tau_{90\%-10\%}$ of the gain after a gain compression is a useful metric. In Fig. 2.18, the dependance of $\tau_{90\%-10\%}$ on pump power and bias current density is displayed. The phase recovery is clearly dominated by the slow process. It shows an effective recovery time of hundreds of picoseconds and cannot be improved considerably by changing the input power or the bias current density. The gain recovery is considerably faster. The associated effective time constant increases with input power, because the gain depletion is much stronger, but it rapidly decreases for higher bias current densities, as the capture efficiency of the QD states becomes larger.

Eventually, the slowest gain recovery time in an SOA, namely the refilling of the carriers from an external current supply τ_c, limits the linear amplifier speed in a data communication system.

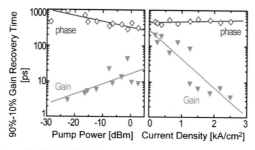

Fig. 2.18: Dependance of effective 90%-10% gain and phase recovery time on input power and bias current density. The phase recovery is dominated by the slow process. The gain recovery is fastest for a low input power and high current densities [J10] and [B1].

From the pump-probe measurement results, we know that in a QD SOA fast and slow processes influence the material gain and the phase dependance simultaneously, but with different magnitudes: The material gain depletes strongly, but recovers fast. In contrast, the phase recovers slowly. Therefore, the effective alpha factor changes with time, Fig. 2.19 [J10]. The alpha factor of the first 10 ps is governed by the refilling of the depleted QD states from the reservoir. This depletion can be described as a SHB process with an alpha factor close to zero. The depleted reservoir states lead to an increased alpha factor ($t > 20$ ps) which remains large until the reservoir states are slowly refilled by external carrier injection.

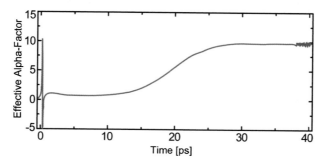

Fig. 2.19: Time-dependent effective alpha factor for an SOA, here for a QD SOA. The dynamic response of the first 10 ps is governed by fast QD refill processes with an alpha factor close to zero. Later on, the depleted reservoir states lead to an alpha factor, which remains at a high level until the reservoir is slowly refilled by externally injected carriers [J10] and [B1].

2.2.7 Modulation Response of an SOA

The performance of an SOA when amplifying intensity-encoded signals with a small modulation depth is of interest, e.g., for radio-over-fiber (RoF) applications. In this section, the optical modulation response of an SOA is introduced and interpreted.

In general, a small-signal model is used to linearize a transfer function for small variations around an operating point. The optical small-signal response of an SOA is defined as $H(f_e)$ with the electrical modulation frequency f_e. The small-signal response describes the behavior of a harmonic small optical power variation $\Delta P_{out}(t)$ around an average optical power P_{out} at the output of the SOA as a function of small optical power variations $\Delta P_{in}(t)$ around an average optical power P_{in} at the SOA input. The frequency-dependent small optical power variations at the SOA output $\Delta P_{out}(f_e)$ and SOA input $\Delta P_{in}(f_e)$ are obtained from the time-dependent changes via the Fourier transform, neglecting terms of higher order. Using the angular frequency ω_e, the electrical phase at the SOA input $\phi_{e,in}$, the electrical phase at the SOA output $\phi_{e,out}$, and the power amplitudes $\Delta P_{in,out}$, we can write

$$H(f_e) = \frac{\Delta P_{out}(f_e)}{\Delta P_{in}(f_e)}, \qquad \tilde{P}_{in}(t) = P_{in} + \Delta P_{in}(t) = P_{in} + \frac{\Delta P_{in}}{2}\sin\left(\omega_e t + \phi_{e,in}\right),$$

$$\tilde{P}_{out}(t) = P_{out} + \Delta P_{out}(t) = P_{out} + \frac{\Delta P_{out}}{2}\sin\left(\omega_e t + \phi_{e,out}\right). \tag{2.55}$$

$H(f_e)$ includes the SOA gain G at the operating point. For RoF applications, we are interested in the SOA's impact on the input signal in terms of the optical modulation depth $m_o(f_e)$ at different SOA operating points. Therefore, we define the modulation response $H_m(f_e)$

$$H_m(f_e) = \frac{m_{o,out}(f_e)}{m_{o,in}(f_e)} = \frac{1}{G}\frac{\Delta P_{out}(f_e)}{\Delta P_{in}(f_e)} = \frac{1}{G}H(f_e), \qquad m_{o,in}(f_e) = \frac{\Delta P_{in}(f_e)}{P_{in}},$$

$$m_{o,out}(f_e) = \frac{\Delta P_{out}(f_e)}{P_{out}}. \tag{2.56}$$

$H_m(f_e)$ is of importance in RoF applications due to the fact that the modulation depth of the RF-signal can be strongly suppressed depending on the SOA operating point and the RF-carrier frequency. In Fig. 2.20 the behavior of an SOA for RF-signals as a function of the optical input power and the RF-carrier frequency is discussed.

First, for low optical SOA input powers $P_{in} \ll P_{in}^{sat}$, the gain G_0 is almost constant, see Fig. 2.20(a). In this linear regime, the output power is in proportion to the input power, independent of the modulation frequency (considering a large SOA gain bandwidth and a constant gain). Following (2.56), the magnitude of the modulation response $H_m(f_e) \to 1$ and the modulation depth at the SOA output is identical to the modulation depth at the SOA input $m_{o,out} = m_{o,in}$.

Second, for large input powers into the SOA $P_{in} > P_{in}^{sat}$ two extreme values for the modulation response of an SOA are found. Considering a modulation frequency of $f_e \to 0$, the SOA dynamics can follow the input power variations. This way, higher input power levels experience less gain than lower input power levels. Thus, the optical modulation depth is reduced: $m_{o,out} < m_{o,in}$ (see Fig. 2.20(b)) and distortions on the signal occur. In contrast, for the case $f_e \to \infty$, all input power levels within the small range of variation experience the same gain G, i.e., the SOA dynamic is too slow to follow the power fluctuations of the input, see Fig. 2.20(c). Then, the modulation depth at the SOA output is identical to the modulation depth at the SOA input $m_{out} = m_{in}$ and $H_m(f_e) \to 1$.

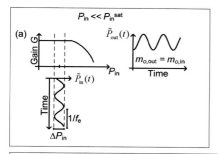

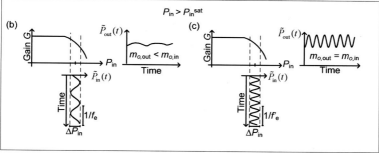

Fig. 2.20: Explanation of the SOA modulation response as a function of the input power and the modulation frequency f_e. (a) If the optical input power is much smaller than the saturation input power, the SOA amplifies the harmonic variations roughly with the same gain independently of the modulation frequency. Therefore, the magnitude of the modulation response $H_m(f_e)$ approaches 1. If the optical input power exceeds the saturation input power, we have to distinguish two cases: (b) The modulation frequency is low so that the SOA dynamic can follow the input power variations. This way, higher input power levels experience less gain than lower input power levels, and the modulation depth is reduced. (c) For high modulation frequencies, the SOA dynamic is too slow to follow the power variations at the input, and all input power levels within the small range of variation experience the same average gain. This way, the modulation depth is maintained and the magnitude of the modulation response $H_m(f_e)$ approaches 1.

Measurement of SOA Small-Signal Response and Results

The modulation response of an SOA can be measured with a lightwave component analyzer (LCA), see Fig. 2.21(a). The LCA is an electrical network analyzer whose electrical ports are connected to components of an optical communication system. The modulation response measurements are performed by sweeping the RF-frequency and thus the optical modulation frequency at the transmitter (T_x) with the help of a Mach-Zehnder Modulator (MZM). The used LCA generates an optical sinusoidal intensity modulation with a modulation depth of approximately 10 % within a frequency range of 0.13 GHz to 20 GHz. This signal is launched into the SOA and the resulting signal is directly detected with a photodiode at the receiver (R_x). The detected signal from the photodiode is analyzed by the LCA in magnitude and phase and compared to the originally transmitted signal as a function of the modulation frequency.

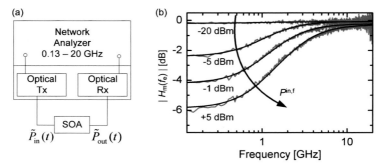

Fig. 2.21: Evaluation of the modulation response of an SOA. (a) Lightwave component analyzer consisting of a network analyzer that is connected to an optical test set for small-signal characterization of SOA. (b) Magnitude of the measured modulation response (red lines) as a function of different SOA input powers. The black curves are fits to the experimental results. The SOA has a G_{f0} of 19.5 dB and an in-fiber saturation input power of −4 dBm.

The magnitude of the modulation response $H_m(f_e)$ in the frequency range of 0.13 GHz to 20 GHz of an SOA as a function of different in-fiber input powers is shown in Fig. 2.21(b). The SOA is operated at $I = 300$ mA at a wavelength of 1300 nm. Note that the fiber-to-fiber gain G_{ff} used to determine the modulation response $H_m(f_e)$ may vary with the optical in-fiber input power $P^{in,f}$ into the SOA.

For low optical input power ($P^{in,f} = -20$ dBm) the SOA is operated in the linear regime which can be seen by a nearly flat magnitude modulation response which is almost reaching 0 dB, thus, no change of the modulation depth is induced by the SOA.

For optical input powers larger than the in-fiber saturation input power ($P_{sat}^{in,f} = -4\,\mathrm{dBm}$) the SOA is operated in the nonlinear regime. Then, for low modulation frequencies the SOA provides different gains for different input powers. This leads to a decrease of the optical modulation depth dependent on the degree of SOA saturation. In contrast, for high modulation frequencies the SOA dynamic is not sufficiently fast to follow the input power variations and all power levels within the modulated optical signal experience the same average FtF gain. Eventually, for large frequencies f_e, the magnitude response reaches 0 dB again.

2.2.8 SOA Nonlinearities

The nonlinear regime of an SOA is reached if the input power exceeds the saturation input power. In this regime, nonlinear effects such as self-gain modulation (SGM), self-phase modulation (SPM), cross-gain modulation (XGM), cross-phase modulation (XPM) and four-wave-mixing (FWM) are observed. These effects are used to build functional devices for different applications as introduced in *Section 2.1.6*. In this section, the nonlinear effects *SGM, SPM, XGM, XPM* and *FWM* are briefly introduced following the description of [20], [62], [101], [113].

- *SGM* in an SOA is the modulation of the gain G at a wavelength λ_1 due to power variations of a signal at the same wavelength λ_1. This is a different way of looking at saturation.

- *SPM* in an SOA is the modulation of the SOA output phase $\Delta\phi$ at a wavelength λ_1 due to power variations of a signal at the same wavelength λ_1. *SPM* and *SGM* are interrelated by the amplitude-to-phase coupling via the alpha factor, see (2.7).

- *XGM* in an SOA is the modulation of the gain G induced by power variations of one signal at a wavelength λ_1 that affects the gain of all other signals with different wavelengths propagating in the same SOA.

- *XPM* in an SOA is the modulation of the output phase $\Delta\phi$ induced by power variations of one signal at a wavelength λ_1 that affects the phase of all other signals with different wavelengths propagating in the same SOA.

- *FWM* in an SOA is the interaction of an input pump signal at an angular frequency ω_1 with a probe signal at an angular frequency ω_2 generating new signals at angular frequencies of $2\omega_1 - \omega_2$ and $2\omega_2 - \omega_1$. This effect occurs due the fact that the pump-wave and the probe-wave induce a gain and refractive index grating within the SOA with a period corresponding to their beat frequency. A third wave which is either from the pump or from the probe signal is modulated by this carrier density (gain) modulation, generating modulation sidebands. In consequence new frequency components are generated. The process requires energy and momentum conservation.

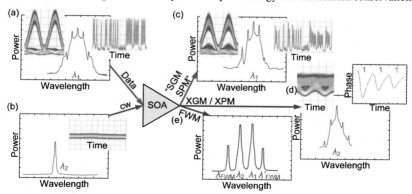

Fig. 2.22: Schematic presenting SOA nonlinearities. (a) Spectrum of input signal, pattern and eye diagram. (b) cw signal and its spectrum. (c) Spectrum and time domain of the signal after *SGM* and *SPM*. (d) Spectrum and eye diagram as well as phase over time (for an initial '111' sequence) of the signal after *XGM* and *XPM*. (e) *FWM* spectrum presenting two input signals and the new frequency components.

Fig. 2.22 shows an example of the nonlinear effects for two input signals into an SOA. An OOK data signal at a wavelength λ_1 (red) is launched into the SOA together with a cw signal at a wavelength λ_2 (blue). In Fig. 2.22(a,b) the signal power as a function of wavelength and the signal power as a function of time for the two input signals are presented. The effects of *SGM* and *SPM* are to be seen in Fig. 2.22(c). The power spectrum centered at the wavelength λ_1 shows an unsymmetrical broadening compared to the input power spectrum from Fig. 2.22(a) due to the generation of new frequency components by *SPM*. The *SGM* effect becomes especially visible in the SOA output power versus time figures (eye diagram and pattern). The OOK data pattern and the eye diagram shows pattern dependent amplification which means that the gain established for a certain bit-slot depends on the previous bit-slots.

The effects of *XGM* and *XPM* are presented in Fig. 2.22(d). The power spectrum around the wavelength λ_2 shows that the modulation from the data signal at the wavelength λ_1 is transferred. The power spectrum further shows an unsymmetrical broadening of the right and the left spectral components due to *XPM*. The *XGM* effect becomes especially visible in the power versus time figure (eye diagram) at the SOA output. The eye diagram for the OOK signal shows that the data pattern has been inverted by means of the *XGM* effect. Pattern dependent amplification causes an eye closure. The phase-over-time inset represents the effect of *XPM* for a bit sequence of '*111*'. Since the *XGM* suffers from bit-pattern effects, the time-dependent phase of the *XPM* effect shows a similar behavior.

The effect of *FWM* is presented in Fig. 2.22(e). The power spectrum shows that due to the input signals at the wavelengths λ_1 and λ_2, new spectral components at λ_{FWM} and λ'_{FWM} (purple) are generated. *FWM* is modulation format transparent and works at highest bit rates. However, it significantly suffers from a limited conversion efficiency, and thus from the achievable low OSNR.

2.3 Quantum–Dot SOA

In addition to the investigations of conventional SOA performed in this work, novel QD SOA are studied for their suitability for linear and nonlinear applications. In this section, promises and accomplishments of these devices are discussed. First the properties of QD SOA are introduced, e.g., their energy-band diagram, the function of the wetting layer, the density of states and the two QD SOA operating regimes. A numerical QD SOA model is presented based on the results [111] and including ASE and phase dependencies. This section extents the already presented theory of conventional bulk or QW SOA.

2.3.1 Properties of QD SOA

In *Section 2.1.3*, it has been shown that the electronic DOS of charge carriers in a semiconductor is significantly modified if the wave function is spatially confined. In a three-dimensional charge carrier confinement the energy levels are quantized in all spatial directions and the energy levels are discrete, see Fig. 2.6. A novel attempt to build SOA is the use of QD-based active regions. Fig. 2.23(a) shows a schematic of an active region of an SOA based on QD. The Stranski-Krastanow (SK) growth mode leads to some variations in size and shape of the dots leading to spread of localization energy within a homogeneously-broadened dot ensemble. In Fig. 2.23(b), a typical lineshape of such a QD ensemble gain spectrum is presented. It is composed of a superposition of numerous Lorentzian single-dot lineshapes. The total gain spectrum is inhomogeneously broadened (typically > 50 meV). In recent years, it has been found (including the results of this work) that the carrier distribution function of a real dot ensemble at room temperature (typically 300 K) is comparable to the one found for the bulk or QW materials [114]. This shows the challenge to use QD SOA for multi-wavelength signal processing in the nonlinear gain regime.

Further, the SK growth causes the formation of a wetting layer of QD material underneath the dots which influences the carrier capture and thus the SOA dynamics. Typically, QD devices employ ensembles of QD with a high density of mid 10^{10}cm^{-2} QD in a single layer. The QD layers are stacked to increase the gain.

Tuning ability of the emission wavelength by the control of the QD size and composition allows the fabrication of devices with greatly expanded gain bandwidth of up to 120 nm [24]. In the future, it is to be expected that attempts will be undergone to extend the SOA gain bandwidth in a large spectral range from 800 nm to 1800 nm within a single QD SOA device.

The carrier confinement into the QD efficiently suppresses carrier diffusion and escape from the QD so that the possibility for uncooled operation increases compared to conventional SOA devices [57]. A disadvantage is the polarization sensitivity of the gain due to the pyramidal shape of the QD. However, the polarization sensitivity can be counteracted by novel and advanced growth and stacking techniques to achieve polarization insensitive operation [43].

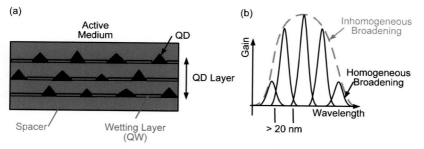

Fig. 2.23: QD active region and gain spectrum. (a) The active region of a QD SOA consists of several stacks of QD layers. Each QD layer comprises QD with different shape and size. The QD are surrounded by a wetting layer which acts as a carrier reservoir. (b) The inhomogeneously broadened gain spectrum of a QD SOA consists of several homogeneously broadened sub-ensembles of QD.

Current injection in a QD is schematically shown in Fig. 2.24(a,b). The energy-band diagram for a single dot and its wetting layer is introduced in Fig. 2.24(a). The primary reason that QD SOA can show very fast gain recovery compared to that of conventional devices is the presence of the wetting layer. The wetting layer has an energy-band diagram of a QW. It can be seen that the carriers are injection from the bulk material into the QW within a capture time τ_2 of around $100...200$ ps, see also Fig. 2.17. The capture into the QD ground state from the surrounding QW (wetting layer) is a fast process of a few ps (τ_1), see also Fig. 2.17. In Fig. 2.24(b) the corresponding DOS of the energy-band diagram from Fig. 2.24(a) is presented, taking into account QD ensembles. Obviously, a QD SOA grown in the SK growth mode combines the DOS of the bulk and QW acting as carrier reservoirs to refill the active QD states.

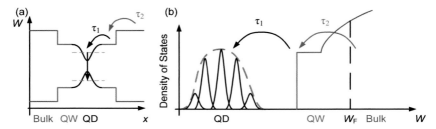

Fig. 2.24: Carrier injection, energy-band diagram and DOS of a QD. (a) Carriers are injected from the bulk material into the QW (wetting layer) by a slow carrier transfer. On the contrary, the QD states are refilled by fast carrier transfer from the QW. (b) DOS for the band energy diagram in (a). The picture has been modified from [115].

The importance of a high carrier injection to observe the fast gain recovery in a QD SOA is discussed in the next subsection.

2.3.2 Two Operating Regions of QD SOA

Two different operating regions can be defined for a QD SOA based on the filling of the wetting layer (WL) and the QD states, namely the regimes of low and high carrier injection, see Fig. 2.25(a)-(h). The energy dependance W of the density of states $D(W)$ for the two regimes is shown in Fig. 2.25(a) and (e), respectively.

In the low carrier injection regime (upper row), the QD states are initially completely filled. After the gain suppression by an input signal, the recovery process takes place at two different time scales, Fig. 2.25(b). Initially, the recovery is fast ($\tau_1 \sim 1$ ps), representing the ultra-fast refilling of the QD states by the low number of available WL carriers. But then, it is significantly slowed down on a time scale of more than 100 ps due to the slow refilling of the WL. This may result in a partially filling of the QD states for, e.g., a long sequence of marks. Thus a significant bit patterning at high bit rates can be observed [116], [117]. Phase effects are limited and introduced primarily by the changes on the WL carrier density, Fig. 2.25(c). In particular, the WL carrier depletion generates red chirped frequencies at the leading edges of a wavelength converted pulse, whereas the trailing edges become blue chirped from the subsequent WL recovery process [118], see Fig. 2.25(d). On the other hand, QD depletion generates a symmetric spectral hole, therefore its phase contribution can be considered negligible [110], [119]. In addition, calculations have shown that plasma-induced phase effects are minor for devices of moderate length [117].

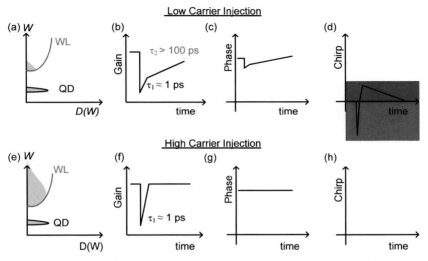

Fig. 2.25: Energy W vs. density of states $D(W)$ and gain as well as phase vs. time for a QD SOA with low (upper row) and high carrier injection (lower row). In the low carrier injection regime, (a) the QD states are initially completely filled. If a sequence of consecutive ones with large input power is launched into the device the QD states will be only partially filled. (b) The gain has an initial ultra-fast recovery (~1 ps) due to the refilling of the QD states from the WL reservoir. Afterwards the recovery slows down (~100 ps) by the necessary refilling of the WL states. (c) Phase effects are minor and primarily introduced by the changes of the WL carrier density. (d) The gain saturation induces red and blue chirped frequency components. In the regime of high carrier injection, (e) the QD states are always fully populated (after a small time interval of the fast refilling). (f) The gain dynamics are dominated by the ultra-fast recovery of the QD states. Neither (g) phase modulation nor (h) chirp occurs, [J7].

So far, state-of-the-art devices operate mostly in the regime of low carrier injection. The current fabrication technology of QD SOA makes it difficult to produce devices which tolerate high injection currents. When increasing the current density, thermal effects can damage the quantum-dot structure. In the near future, however, it is likely that devices with higher tolerance of large injection currents can be fabricated. In that operating regime, the carrier population in the QD states and in the WL are effectively increased, Fig. 2.25(e). The ultra-fast dynamics (~1 ps) of the QD states dominate, and pattern-effect free operation is expected as long as bit-periods are larger than the respective QD refilling time, Fig. 2.25(f). On the other hand, since the WL is not depleted, no phase modulation occurs, Fig. 2.25(g) and the induced chirp is zero Fig. 2.25(h).

2.3.3 Numerical Rate–Equation Model for QD SOA

The numerical rate-equation model introduced in this subsection is a modification of the QD SOA modeling approach proposed by Uskov [111], [119]. Compared to other approaches [23], [120] that perform a detailed evaluation of each dot group (ensemble), this model takes into account only the dynamics of the QD that are in resonance with the input optical signals. The remaining non-resonant groups are treated, along with the WL, as a part of the same carrier reservoir. This approach has significant advantages in terms of computational efficiency and enables a more intuitive explanation of the dynamic performance of the device. In the model also the phase dynamics as well as the accumulation of the generated ASE noise along the active waveguide are included.

This model is used to describe the amplification of advanced modulation format data signals with SOA in *Section 4.4.1*, and the wavelength conversion performance of QD SOA in the two operating regimes in *Section 5.3*.

The evolution of the photon density S of two input channels along the propagation direction z in the device, namely the modulating pump S_{pump} and the cw probe S_{probe}, are governed by the propagation equation

$$\frac{\mathrm{d}S_k}{\mathrm{d}z} = g_k S_k, \qquad k = \text{pump,probe}. \tag{2.57}$$

This relation is defined in a retarded time frame that moves with the photons at group velocity v_g. The parameter g_k represents the modal net gain of the pump or the probe signal. The confinement factor Γ is defined here as the ratio of the total QD layer thickness to the height of the active waveguide. Along z, both signals interact through the cross-saturation of their respective modal net gains g_k. This process depends on the carrier depletion mechanisms of both the resonant QD group and the reservoir. If at each infinitesimal section of the waveguide the total photon density $S_{total} = S_{pump} + S_{probe}$ is relatively low, the corresponding carrier depletion mechanisms for QD and WL can be decoupled. In addition, if the frequency difference of the two input signals is much less than the homogeneous linewidth of the resonant QD, then the net modal gains of pump and probe signal coincide [116], [119]

$$g \equiv g_{pump} \approx g_{probe} \approx \frac{g_{lin}}{1 + \varepsilon_{SHB} \cdot S_{total}}, \tag{2.58}$$

where $\varepsilon_{SHB} = \tau_1 \sigma_{res} v_g$ is the nonlinear gain coefficient for SHB with the QD refilling time τ_1 and the resonant cross-section of carrier-photon interaction σ_{res} [119]. The nonlinear gain coefficient ε_{SHB} describes the virtually instantaneous suppression of the modal net gain g due to the very fast stimulated recombination of carriers at the resonant dot group. In Eq. (2.58) the linear part of the modal net gain g_{lin} is represented, which depends on the instantaneous value of the total carrier density (volumetric) in WL and QD, and is only indirectly and more slowly affected via the wetting layer refilling time by the total photon density S_{total}. The total carrier density N is governed by the following rate equation

$$\frac{dN}{dt} = -\frac{N - N_{J}}{\tau_{c}} - \frac{\upsilon_{g} g_{lin} S_{total}}{1 + \varepsilon_{SHB} \cdot S_{total}}.$$ (2.59)

Here, $N_{J} = \eta_{i} J \tau_{c} / (e t_{wg})$ is the pump strength parameter, J is the current density, η_{i} the injection efficiency of the carriers in the active region, t_{wg} is the effective optical thickness of the SOA waveguide, τ_{c} is the effective carrier lifetime (associated with the slow refilling of the wetting layer τ_{2}), and e the electron charge. The dependance of the linear gain on the instantaneous value of the total carrier density is defined through the QD filling factor $W_{o}(N) = 2p_{o} - 1$, where p_{o} is the occupation probability of the resonant QD. $W_{o}(N)$ indicates the QD filling in thermal quasi-equilibrium between the resonant QD and the reservoir. Following the analysis of [119] we may write

$$g_{lin} = g_{max} \cdot W_{o}(N) = l\frac{2N_{D}}{t_{wg}}\sigma_{eff} \cdot \underbrace{(2p_{o} - 1)}_{W_{o}(N)},$$ (2.60)

$$\exp\left[\frac{N - \dfrac{2N_{D}p_{o}l}{t_{wg}}}{N_{wL}}\right] = 1 + \exp\left[-\frac{W_{bind}}{kT}\right] \cdot \frac{p_{o}}{1 - p_{o}}.$$ (2.61)

Here, N_{D} is the density of the QD per unit area in each layer, l is the number of QD-layers, $\sigma_{eff} = \sigma_{res}(\pi \ln 2)^{-1/2}\gamma_{hom} / \gamma_{inhom}$ is the effective cross-section for the photon-carrier interaction, γ_{hom} represents the homogeneous linewidth of the resonant QD and γ_{inhom} is the inhomogeneous linewidth of the QD ensemble. Eq. (2.61) relates the occupation probability p_{o} with the available carrier density N of the reservoir in the condition of thermal quasi-equilibrium. $N_{wL} = mklT / (\pi \hbar^{2} t_{wg})$ represents the 3D effective density of states for the WL, where m is the effective carrier mass (identical for electrons and holes), and T is the temperature of the QD structure. Furthermore, W_{bind} is the average binding energy of the QD ensemble, defined relative to the position of the WL edge. Through numerical solution of (2.61), the dependency of the QD filling factor $W_{o}(N)$ on the carrier density N can be derived. The result is illustrated in Fig. 2.26. The horizontal axis is normalized by the volumetric density of the degenerated QD states $2N_{D}l / t_{wg}$. For low carrier density values (low carrier injection regime) the QD filling is incomplete ($W_{o} < 1$). This means that the reservoir has not enough carriers to fill the ground state levels of the resonant QD, and its slow refilling processes will govern the dynamic behavior of the device.

When the total carrier density is increased by increasing the injection current, the QD filling factor approaches the value of $W_{o,max} = 1$, designating a regime of complete population inversion (high carrier injection regime). In that case, the linear gain becomes constant and acquires a maximum value of $g_{max} = 2N_{D}l \sigma_{eff} / t_{wg}$. Furthermore, the modal net gain response to the changes of the input photon density occurs instantaneously, see (2.58), enabling pattern-effect free data processing for both linear and nonlinear operation [119]. Practically, this is achieved when the normalized carrier density is beyond a characteristic value N_{c} (see Fig. 2.26). Actually, N_{c} depends on characteristic material and structural properties of the QD SOA such as temperature, effective carrier mass, energy difference between WL and QD states as well as dot density [121]. For high effective carrier masses and a low energy difference, there is a strong coupling between the energy states of WL and QD. The filling of the QD is then strongly influenced by the available WL states and requires higher reservoir carrier densities for the same level of inversion. The density of QD plays also an important role. Reducing the number of QD in the device increases in effect the number of reservoir carriers that are available to fill the empty QD states. Therefore, complete QD filling becomes feasible at significantly lower injection currents.

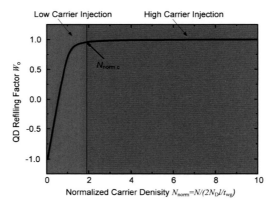

Fig. 2.26: QD refilling factor W_o as a function of the normalized carrier density N_{norm}. Full QD filling and thus complete population inversion is achieved when $N_{norm} > N_c$, [J7].

Detailed knowledge on the phase dynamics is also required. Changes in the carrier density also change the refractive index besides affecting the gain. This interplay is well described by the Kramers-Kronig integral expressions. However, an easier to handle tool for characterizing the nonlinear phase shift observed in connection with absorption or gain in nonlinear media is the alpha factor α_H [112]. If we assume that the amplifier is operating close to the gain peak, so that a symmetrical spectral hole does not contribute to the nonlinear index change, and if we ignore plasma effects, then a phase-amplitude coupling coefficient may be defined for the changes of the total carrier density only.

The phase ϕ of an optical field propagating through the SOA waveguide evolves according to

$$\frac{\partial \phi}{\partial z} = \frac{1}{2}\alpha_{\mathrm{H}} \cdot \left[g_{\mathrm{lin}}\left(N\right) - g_{\mathrm{lin}}\left(N_{\mathrm{J}}\right) \right]. \tag{2.62}$$

The accumulation of the generated ASE noise along the active waveguide has also been taken into account. Details on the ASE noise modeling are presented in [J7].

The simulations are performed with an active QD SOA region which is subdivided in 20 finite elements. In these segments the rate equations are solved. The solution (output field) of the rate equations in a segment is always passed to the subsequent segment at which it acts as the input field.

3 SOA Samples and Characterization Results

The performance of different semiconductor optical amplifier (SOA) devices is experimentally investigated in this thesis for network applications such as linear amplification (*Chapter 4*) and nonlinear signal processing (*Chapter 5*). This chapter connects the SOA parameters (*Section 2.2*) with the above mentioned chapters. Here, the steady-state and the dynamic properties of the used SOA are presented. The steady-state characterization comprises measurements of the fiber-to-fiber (FtF) gain G_{ff}, the polarization dependent gain (PDG), the 3 dB gain bandwidth B_G, the FtF noise figure NF_{ff}, the 3 dB saturation input power $P_{\text{sat}}^{\text{in,f}}$, and the effective alpha factor α_{eff}. The dynamic characterization was performed using pump-probe measurements to derive the 90 %-10 % gain recovery time $\tau_{90\%-10\%}$.

In *Section 3.1*, the SOA samples used in this work are introduced. In *Section 3.2*, the characterization results of the samples are reported.

3.1 SOA Samples

In this work, the linear and nonlinear properties of SOA are studied. The investigated SOA samples may be distinguished by the dimensionality of the electronic system of their active region, by the gain peak wavelength λ_{Peak}, by the geometrical structure, their length L, and their active waveguide width w. Quantum-dot (QD) SOA and quantum-well (QW) / bulk SOA are used with gain peak wavelengths close to 1.30 µm and 1.55 µm, respectively. They are formed with ridge waveguides or with buried heterostructures. Table 3.1 summarizes the SOA parameters.

No.	Name	Active Region	Structure	Length L [mm]	Width w [µm]	Application
SOA 1	TUB-DO957	QD	Ridge	4	2	LA
SOA 2	TUB-DO957	QD	Ridge	4	4	LA
SOA 3	TUB-DO1202	QD	Ridge	2	4	NLA
SOA 4	III-V-Lab A	QD	Buried	1	1.75	LA
SOA 5	III-V-Lab B	QD	Buried	2	1.5	NLA
SOA 6	III-V-Lab C	Bulk	Buried	0.7	1.5	LA
SOA 7	Company A	MQW	n/a	n/a	n/a	LA

Table 3.1: Overview of the SOA used in the experimental investigations within this work. The abbreviation LA means that the SOA are applied for linear applications whereas NLA means nonlinear applications, respectively. n/a stands for not applicable.

SOA 1, 2 and 3 are QD SOA which are fabricated by the Technical University of Berlin (TUB, group of Prof. Dr. D. Bimberg). The samples are grown by molecular beam epitaxy (MBE), and consist of 10 layers of In(Ga)As/GaAs QD. Each InAs QD layer was formed by self-organized growth in the Stranski-Krastanow mode following growth interruption after deposition of 2.5 monolayers of InAs directly on a GaAs matrix [51], see *Section 2.1.3*.

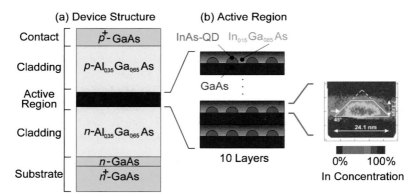

Fig. 3.1: Epitaxial layer structure of the QD SOA (SOA 1, 2 and 3). (a) The device structure shows that on a GaAs substrate an n-AlGaAs cladding layer is grown. Then, using the Stranski-Krastanow growth mode the active region containing the QD layers is generated. Finally, a p-AlGaAs cladding and the contact layer is grown. (b) The active region consists of 10 InAs-QD layers. The QD are covered by an InGaAs-QW. The QD layers are separated by GaAs spacer. The In concentration of a single QD is also presented. The figure is modified from [115].

Fig. 3.1(a) shows the epitaxial structure of the samples SOA 1, 2 and 3. The QD are grown on an n^+-doped GaAs (001) substrate. The active region consists of 10 stacked layers of self-assembled InAs quantum dots, each covered with a 5 nm thick $In_{0.15}Ga_{0.85}As$ QW (dot-in-a-well-structure, see Fig. 3.1(b)) in order to shift the emission wavelength to 1.30 μm. The QD are approximately 5 nm in height and 15...20 nm in lateral extension. Strain relaxation is provided by a 33 nm thick GaAs spacer layer between the QD layers. The QD density per layer is approximately $3...5 \times 10^{10} cm^{-2}$. The active region is surrounded by 1.5 μm thick p(Be)- and n(Si)-doped $Al_{0.35}Ga_{0.65}As$ cladding layers above and below [113], [115], [122] to provide vertical waveguiding. On top a contact layer for current supply is grown. Additionally, it should be mentioned that the SOA 3 has a p(Be)-doped active region. In this case, a 10 nm thick part of the spacer layer which is 9 nm below the QD layer is p-doped ($5 \times 10^{17} cm^{-3}$). This leads to a prefilling of the QD hole states.

The SOA are processed to shallow etched ridge waveguide structures with a length L of 4 mm (SOA 1 and SOA 2) and 2 mm (SOA 3) and waveguide widths w of 2 μm (SOA 1) and 4 μm (SOA 2 and SOA 3), respectively. Shallow etching is understood here, as etching the cladding layers up to the active layer containing the QD. This process provides ridge waveguides which offer weak index guiding of the optical mode. The waveguides are tilted by 8° normal to the facets to minimize reflections at facets. No anti-reflection coating is used.

SOA 4 and SOA 5 are also QD SOA. The devices are fabricated and provided by Alcatel Thalès III-V-Lab, a joint lab of Bell Labs and Thales Research & Technology. The samples are grown by MBE on (100) InP wafers, using the self-organized Stranski–Krastanow growth mode. The QD are generated based on strain relaxation occurring in a very thin ($\sim$1 nm) highly strained InAs layer deposited on an InGaAsP layer. The mismatch between the material lattice constants, of typically 4 % in this material system, induces the QD formation. The typical width of these dots is about 15...20 nm. The length of the dots depends on the growth conditions, ranging between 40...300 nm. The surface density of QD is about $1...4 \times 10^{10} \mathrm{cm}^{-2}$ [49]. The samples SOA 4 and SOA 5 consist of 6 layers of QD each. The SOA are processed to buried heterostructures with a length L of 1 mm (SOA 4) and 2 mm (SOA 5) and waveguide widths w of 1.75 μm (SOA 4) and 1.5 μm (SOA 5), respectively. The waveguides are tilted normal to the facets to minimize reflections at the facets, and an anti-reflection coating is used.

SOA 6 is a bulk SOA which is grown by MBE. This sample is also fabricated and provided by Alcatel Thalès III-V-Lab. SOA 6 is a buried heterostructure with a length of 0.7 mm and a waveguide width of 1.5 μm.

SOA 7 is a commercial available amplifier which is based on the InP multi-quantum well (MQW) technology.

3.2 Steady–State and Dynamic Characterization Results

The SOA 1...7 are characterized in steady-state and dynamic measurements. The measurement results for λ_{Peak}, unsaturated fiber-to-fiber gain G_{f0}, B_G, PDG, $P_{\mathrm{sat}}^{\mathrm{in,f}}$, NF$_{\mathrm{ff}}$, $\tau_{90\%-10\%}$ as well as α_{eff} of the devices are presented in Table 3.2. The used current density J is also shown.

The FtF gain, the 3 dB gain bandwidth, the polarization dependent gain, and the noise figure values are obtained using a setup similar to Fig. 2.12 and the definitions introduced in *Section 2.2.2* and *Section 2.2.3*, respectively. Once, the FtF gain as a function of the input power has been measured, the 3 dB in-fiber (power measured in the fiber) saturation input power is determined accordingly to *Section 2.2.4*. The SOA dynamics are measured in pump-probe experiments with the setup introduced in Fig. 2.16 and the definitions presented in *Section 2.2.5* are applied. The effective alpha-factor values are obtained either from pump-probe measurements as introduced in *Section 2.2.6*, or they are evaluated in simulations using the numerical QD SOA model from *Section 2.3.3*.

SOA 1, 2 and 3 are amplifiers with a gain peak wavelength of about 1295 nm at an operating temperature of 20 °C. The unsaturated FtF gain G_{f0} of SOA 1 is 15.5 dB, 17 dB of SOA 2 and 5 dB of SOA 3, respectively. The 3 dB gain bandwidth of the samples is around 35 nm. The PDG is about 10 dB due to the nature of the QD, see *Section 2.3*. The saturation input powers of the devices SOA 1...SOA 3 are between −7 dBm and 10 dBm. The reason is the strong dependance of the saturation input power on the unsaturated FtF gain G_{f0} and the supplied current density J. The used bias current I is 490 mA for the experiments with SOA 1, 300 mA for the measurements with SOA 2 and 600 mA for the experiments with SOA 3. The noise figure of the devices is around 10 dB.

SOA	λ_{Peak}	G_{f0} @ λ_{Peak}	B_G	PDG	$P_{sat}^{in,f}$	NF_{ff}	$\tau_{90\%-10\%}$	α_{eff}	J
	[nm]	[dB]	[nm]	[dB]	[dBm]	[dB]	[ps]		[kA/cm^2]
1	1295	15.5	35	10	-2	10	< 10	0-10	6
2	1295	17	35	10	-7	10	10-20	0-5	2
3	1295	5	n/a	n/a	10	n/a	< 10	n/a	7.5
4	1530	18.5	60	5	-9	8	n/a	2^*	10
5	1525	20	70	5	-12.5	10	n/a	2^*	8.6
6	1530	18.5	60	1	-8	7.5	n/a	4^*	9.4
7	1280	19.5	60	1.5	-4	7	n/a	n/a	n/a

Table 3.2: Overview of the measured SOA device performance for the SOA which have been introduced in Table 3.1. The performance of the SOA 1…SOA 7 are presented in terms gain peak wavelength, unsaturated FtF gain at the gain peak wavelength G_{f0}, 3 dB gain bandwidth B_G, polarization dependent gain (PDG), saturation input power $P_{sat}^{in,f}$, FtF noise figure (NF_{ff}), 90 %-10 % gain recovery time $\tau_{90\%-10\%}$, the effective alpha factor α_{eff} and the used current density J. (*) indicates that the values are obtained from simulation of these devices, and n/a means not applicable.

The QD SOA have a fast and a slow carrier refilling time, see *Section 2.3*. The fast time constant τ_1 is in the order of 1…10 ps and is due to a quick refilling of the QD by the wetting layer states. The slower time constant τ_2 in the order of 100…200 ps and is due to the refilling of the reservoir states [J10], see Fig. 2.17. The gain recovery is dominated by the fast time constant, see Fig. 2.18. A $\tau_{90\%-10\%}$ gain recovery time in the order of 10…20 ps is measured. On the contrary, the phase response is dominated by the slow component. Thus, the effective alpha factor strongly changes with time under a large signal modulation. The effective alpha factor of the first 10 ps after a pump pulse with an input power exceeding the saturation input power is governed by the refilling of the depleted QD states from the reservoir. This depletion can be described as a spectral-hole-burning (SHB) process with an alpha factor close to zero. The depleted reservoir states lead to an alpha factor increase ($t > 20$ ps) in the range of 5…10 which remains large until the reservoir states are slowly refilled by external carrier injection.

SOA 1 and SOA 2 are used for the measurement of the input power dynamic range (IPDR) for on-off-keying (OOK) data signals, see *Section 4.3*. SOA 2 is additionally used for the study of the SOA reach extender for next-generation passive optical networks, and the orthogonal frequency division multiplexing (OFDM) recirculating fiber-loop experiments, see *Section 4.6*. SOA 2 is also used to study device guidelines for largest IPDR with moderate gain, see *Section 4.3.7*, and for the dynamic range in radio-over-fiber (RoF) systems, see *Section 4.5.1*. SOA 3 is used to perform high-speed all-optical wavelength conversion of an 80 Gbit/s OOK data signal. This measurement has been performed at the Heinrich-Hertz-

Institute (HHI) in Berlin, where also the steady-state characterization has been conducted. The measurement results are presented in *Section 5.3.2.3*.

SOA 4, 5 and 6 are amplifiers with a gain peak wavelength around 1530 nm at an operating temperature of 20 °C. The unsaturated FtF gain is 18.5 dB of SOA 4, 20 dB of SOA 5 and 18.5 dB of SOA 6, respectively. The 3 dB gain bandwidth of the samples is around 60…70 nm. The PDG of the QD SOA samples 4 and 5 is about 5 dB. On the contrary, the PDG of the bulk SOA (SOA 6) is only 1 dB. The in-fiber saturation input power of the devices SOA 4…SOA 6 varies between −12.5 dBm to −8 dBm. The used bias current is 150 mA for the experiments with SOA 4, 300 mA for the measurements with SOA 5 and 100 mA for the experiments with SOA 6. The noise figure of the devices is between 7.5…10 dB. The dynamic behavior of the SOA 4…SOA 6 is measured with a frequency resolved electro-absorption gating (FREAG) technique based on linear spectrograms [123], see *Appendix A.7*. The resolution of this measurement technique is limited by the used input pulse width to about 10 ps, which only allows the observation of the slow carrier refilling time in the order of 100 ps. The fast gain and phase recovery times of the SOA 4…6 are obtained from device simulations using SOA models, see *Section 2.3.3*. A fast carrier refilling time τ_1 for the QD SOA of 1.25 ps for SOA 4 as well as of 3 ps for SOA 5 is calculated. Further, the effective alpha factors of the SOA 4…SOA 6 are obtained from simulations. The alpha factor is 2 for SOA 4, 2.4 for SOA 5 and 3.5 for SOA 6, see *Section 4.4.1.3* and *Section 5.3.1*.

SOA 4 and SOA 6 are used for the measurements of the IPDR for advanced optical modulation format signals, see *Section 4.4*. SOA 5 is used for the experimental realization of wavelength converters at 10 Gbit/s and 40 Gbit/s RZ OOK, which are presented in *Section 5.3.2*.

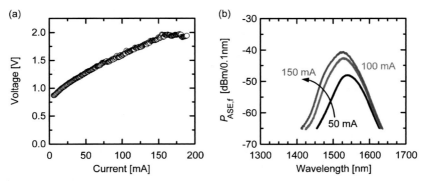

Fig. 3.2: Steady-state characterization results of SOA 4. (a) Forward voltage as a function of current. (b) ASE output power as a function of wavelength for different injection currents.

SOA 7 has a gain peak wavelength of about 1280 nm at an operating temperature of 25 °C. The unsaturated FtF gain is 19.5 dB, the PDG is 1.5 dB and the 3 dB gain bandwidth of the sample is around 60 nm. The in-fiber saturation input power of the device is −4 dBm and the noise figure is around 7 dB.

SOA 7 is used for the amplification of data-burst signals as occurring in an access network, see *Section 4.6.2*.

SOA4 (QD SOA, III-V-Lab): Example of Characterization Results

In Fig. 3.2, exemplary the steady-state characterization results for the series resistance and the amplified spontaneous emission (ASE) power spectrum measured at SOA 4 are presented. Fig. 3.2(a) shows the forward voltage as a function of current. The series resistance is about 6.9 Ω. Fig. 3.2(b) presents the ASE power $P_{ASE,f}$ in a resolution bandwidth of 0.1 nm (measured in a fiber) as a function of wavelength for various bias currents ranging from $I = 50$ mA to 150 mA. The gain peak shifts to smaller wavelengths if the current is increased. At low injection currents, the QD ground state is already fully occupied. Thus at higher injection currents, the QD excited state is significantly filled up with carriers which cause a blue shift of the spectrum. The temperature of the SOA 4 is set to 20 °C for the measurements.

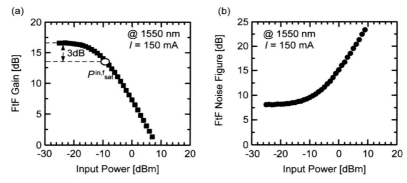

Fig. 3.3: FtF gain and FtF noise figure as function of input power for SOA 4. (a) FtF gain as a function of input power. The 3 dB saturation input power is indicated by a circle. (b) FtF noise figure as a function of input power. The best NF$_{ff}$ of 8 dB is obtained for low input power levels.

The FtF gain G_{ff} as a function of SOA input power $P^{in,f}$ for a current of 150 mA at a wavelength of 1550 nm is shown in Fig. 3.3(a). The saturation input power is indicated by a circle (○). The FtF noise figure NF$_{ff}$ as a function of input power at a wavelength of 1550 nm and a current of 150 mA is shown in Fig. 3.3(b). NF$_{ff}$ has a value of 8 dB as long as the input power level is not exceeding the saturation input power. For higher input power levels the noise figure increases due to gain suppression.

The FtF gain as well as the FtF noise figure as a function of wavelength is shown in Fig. 3.4(a) and (b). The measurements are performed with an input power of −20 dBm with a current of 150 mA. The 3 dB gain bandwidth of 60 nm is indicated. NF_{ff} increases from 8 dB to around 10 dB if the gain decreases.

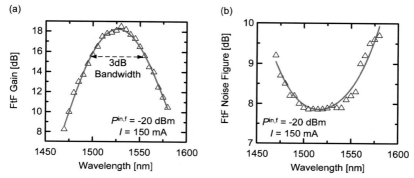

Fig. 3.4: Gain and noise figure as a function of wavelength for SOA 4. (a) The FtF gain as function of wavelength is shown, and the 3 dB gain bandwidth is indicated. (b) The FtF noise figure as a function of wavelength. NF_{ff} rises at the edges of the gain spectrum due to the decreasing gain.

Fig. 3.5(a) shows the phase response of the SOA 4 to an 8 ps wide impulse at a wavelength of 1557 nm with an average input power of +7 dBm. The phase change is measured by means of cross-phase modulation (XPM) onto a continuous wave (cw) signal at a wavelength of 1554 nm. Fig. 3.5(b) displays the peak-to-peak (PtP) phase changes as a function of the channel input power. The average input power of the cw and data channels is always adjusted to be equal, defining the channel input power. The phase changes are measured using the FREAG technique.

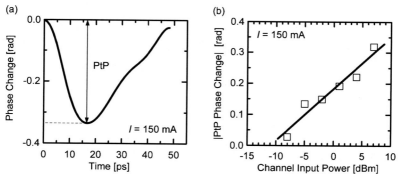

Fig. 3.5: Phase response and peak-to-peak (PtP) phase changes for SOA 4. (a) Phase response to an 8 ps wide impulse. (b) PtP phase changes as a function of the channel input power (always the same cw and data input power levels are used).

This chapter provided a summary of the device specifications, designs and characterization results. Further details on the steady-state characterization as well as on the dynamic characterization are presented in the respective chapters.

4 Linear SOA and Network Applications

Next-generation high-speed all-optical networks will use semiconductor optical amplifiers (SOA) to overcome losses induced by the huge amount of intelligent network components, the long reach and the high power split ratio. Especially the convergence of metro and access networks requires inexpensive optical amplifiers offering large input power dynamics around the wavelength of 1.3 µm and 1.55 µm, respectively. Here, SOA have attracted much interest in recent years due to their ability to amplify data signals over a wide wavelength range.

As outlined in *Section 1.4*, in next-generation optical networks, SOA will need to cope with on-off-keying (OOK) signals, advanced modulation format signals, radio-over-fiber (RoF) signals, e.g., encoded with an intensity-modulated orthogonal frequency-division-multiplexing (OFDM). The question though is how well SOA can amplify such signals and which parameters matter in the selection of an SOA.

An ideal SOA amplifies symbols with both small and large amplitudes alike. This presumably requires an SOA with a large input power dynamic range (IPDR). The IPDR defines the power range in which error-free amplification can be achieved. A high upper limit of the IPDR is obtained when the SOA is linear which means that a constant gain over a large range of input powers can be established. Thus, the SOA has a large saturation input power $P_{\text{sat}}^{\text{in}}$ such that it can cope with largest amplitudes. On the contrary, a good lower limit of the IPDR is obtained when the SOA has a low noise figure such that it can deal with weak amplitudes. Successful attempts towards increasing the linear operating range for OOK data signals introduced gain-clamped SOA or holding beam techniques [102], [124]-[127]. Other approaches exploit special filter arrangements [128], [129] in order to mitigate bit-pattern effects of an SOA when operated in its nonlinear gain regime.

In this thesis, the IPDR of SOA is studied for OOK, advanced modulation formats, RoF signals as well as for multiplexing techniques such as wavelength-division-multiplexing (WDM) and OFDM. Here, the approach is based on the optimization of the SOA devices rather than using schemes to mitigate bit-pattern effects. The SOA parameters introduced in *Section 2.2* are used to derive device design guidelines for a large IPDR. Then, such linear SOA are chosen, i.e., quantum-dot (QD) SOA, to evaluate the IPDR experimentally. The experiments are predicted by simulations based on the QD SOA model from *Section 2.3* and based on the conventional SOA model from *Section 2.2*. Finally, network applications of such linear SOA having large IPDR are predicted and demonstrated experimentally.

In this chapter, the performance of low noise figure and linear SOA is investigated.

In *Section 4.1*, SOA parameters such as gain, noise figure, saturation power, SOA dynamics as well as the alpha factor are discussed in terms of enlarging the region of almost constant gain, i.e., the linear SOA gain regime. Since each SOA adds noise to the signal, the importance of the noise figure is addressed.

The content of this section has been published by the author in [B1] and [J2]. Minor changes have been done to adjust the notations of variables.

In *Section 4.2*, limitations for signal amplification using SOA are outlined, exemplary, for the OOK and phase-shift keying (PSK) modulation formats.

The content of this section has been published by the author in [B1], [C4] and [J1]. Minor changes have been done to adjust the notations of variables and figure positions.

In *Section 4.3*, the IPDR is defined. SOA design guidelines for a large IPDR exceeding 35 dB with moderate gain > 15 dB are reported. Measurements of the IPDR are performed for 2.5 Gbit/s, 10 Gbit/s and 40 Gbit/s OOK data signals with one and two wavelength channels.

Part of the content of this section has been published by the author in [J2], [C13], [C34] and [B1]. Minor changes have been done to adjust notations of variables and figure positions.

In *Section 4.4*, the IPDR for advanced optical modulation format signals are discussed. Experiments with differentially quadrature phase-encoded (DQPSK) data signals with one and with two wavelength channels are performed at 56 Gbit/s. Also non-differentially phase-encoded signals are used to identify the IPDR for 20 Gbit/s binary PSK (BPSK), 40 Gbit/s quadrature phase-shift keying (QPSK) and 80 Gbit/s 16-ary quadrature amplitude modulation (16QAM). Simulation of various modulation formats supports the experimental findings.

Part of the content of the *Subsection 4.4.1* has been published by the author in [J1]. Minor changes have been done to adjust the notations of variables and figure positions. *Subsection 4.4.1.4* has been added containing material about the upper limit and lower limit of the IPDR as function of symbol rate and alpha factor. The content of the *Subsection 4.4.2* has been published by the author in [J1] and [C4]. The content of the *Subsection 4.4.3* has been published by the author in [J6], [C19] and [B1]. Minor changes have always been done to adjust the notations of variables and figure positions.

In *Section 4.5*, the IPDR and the electrical power dynamic range (EPDR) are studied for RoF signals. Experiments with QPSK and 16QAM have been performed with 2 GHz and 5 GHz RoF signals at a symbol rate of 10 MBd.

The content of this section has been published by the author in [C9]. Minor changes have been done to adjust the notations of variables and figure positions.

In *Section 4.6*, the application of linear SOA as reach-extender amplifiers in access networks as well as its performance under burst-mode operation is demonstrated. Further, a cascade of linear QD SOA in a recirculating-loop for OFDM and OOK data signals at a bit rate of 10 Gbit/s is tested to investigate their behavior as loss compensating elements in, e.g., reconfigurable optical add-drop multiplexers (ROADM).

The content of the *Subsection 4.6.1* has been published by the author in [C25]. Minor changes have been done to adjust the notations of variables and figure positions.

4.1   Linear SOA Parameters for Network Applications

Reach-extender amplifier and network elements such as ROADM require linear SOA. In order to get linear amplification across the largest possible spectral and input power range, parameters such as gain G, gain bandwidth B_G, polarization dependent gain (PDG), noise figure (NF), saturation input power P_{in}^{sat}, SOA dynamics as well as the alpha factor α_H need to be optimized. In the following, the parameters (introduced *in Chapter 2, Section 2.2*) which are desirable for linear operation are discussed.

Important Parameters of a Linear SOA for Metro and Access Networks

The SOA parameters strongly depend on the position of the SOA in the network and the network topology itself. A possible configuration is to place the SOA as an in-line amplifier in the network, see Fig. 1.4. Since SOA need to be operated below the saturation input power to avoid bit-pattern effects this position in the network requires the use of a moderate gain and a moderate saturation output power, respectively.

Thus, an unsaturated *fiber-to-fiber (FtF) gain* G_{f0} exceeding 10 dB is ordinarily sufficient to extend the power budget of the network enabling an increased reach from, e.g., 20 km to 60 km, an increased split ratio in a passive optical network (PON) from, e.g., 1:16 to 1:128 or the compensation of insertion loss of network elements, e.g., ROADM, see *Section 1.4*. SOA offer an amplification bandwidth in the range of 50 nm... 80 nm which can be tuned to any range from 1200 nm to 1600 nm by adjusting the material systems (e.g., the composition). Such a gain bandwidth B_G is sufficient for amplification of several data signals which can have spacing in the range from 20 nm to below 1 nm. The PDG has to be below 0.5 dB across the entire SOA bandwidth.

The *FtF noise figure* (NF$_{ff}$) should be as small as possible for amplifying the usually low input power levels into an in-line amplifier. NF$_{ff}$ around 5 to 6 dB is achievable today in commercial devices.

The *saturation input power* P_{sat}^{in} should be as high as possible. This way saturation of the amplifier can be avoided as well as patterning and inter-channel cross-talk. A large saturation input power is of particular importance for an upstream reach-extender amplifier, *Section 1.2.1*, where data signals from the customer locations are sent to the central office (CO). Due to the distance variations of the customer locations, the power at the SOA input varies. For future networks, input power variations in the order of > 40 dB are envisaged. Therefore, an SOA requires a large IPDR to cover power variations due to these distance variations of customer locations.

The *90 %-10 % gain recovery time* is of limited interest since the SOA is operated in the linear gain regime to avoid gain saturation. However, it should be mentioned that the gain dynamic of an SOA has influence on the saturation power.

Lastly, a small *alpha factor* has advantages as well. It guarantees that there are little if no phase variations upon a change in the gain. This way a transition from one point in the constellation diagram to another will not be accompanied by phase changes. This is of particular importance for networks with PSK modulation formats since PSK encodes the information on the phase only, but amplitude transitions from one symbol to the other still might go through the constellation zero.

Influence of Dimensionality of the Electronic System on the Linearity of an SOA

In the following, the parameters are compared for the mature bulk SOA and QW SOA technologies, along with the relatively novel QD SOA. This is done from a theoretical point of view as based on the equations given in the *Chapter 2*, but also relying on experimental studies performed in recent years [21], [23], [50], [111], [112].

The *single-pass gain* G of both bulk and QD SOA can be made comparable, though their lengths, however, are quite different. This can be understood from the definition of the gain G (2.18) and the net modal gain g (2.34),

$$G = \exp(gL) = \exp\left[\left(\Gamma g_m - \alpha_{int}\right)L\right] = \exp\left[a\left(N - N_t\right)L\right]. \qquad (4.1)$$

Thus, the gain G of an SOA depends on the carrier density N, the differential gain a, the amplifier length L and the confinement factor Γ. While quite a few sources state that the differential gain of the different devices is comparable with $a \sim 2 \times 10^{-20}\ \mathrm{m}^2$ [23], [130] the confinement factor is not. The confinement factor of a QD SOA is in the order of 1 % per layer [58] and in the order of 20 %...50 % in bulk and QW SOA [58]. However, the reduction in gain by the low confinement factor is typically compensated by stacking QD layers (5...15 layers are typical), and by making QD SOA devices longer. Typically QW and bulk SOA lengths are in the range 0.5...1 mm and thus have about ~ 25 % of the length of a QD SOA (2...4 mm).

One disadvantage of QD SOA is that they show large PDG of up to 10 dB. However, this PDG can be eliminated with a polarization diversity scheme. On the contrary, an advantage of QD SOA is the large achievable gain bandwidth B_G exceeding 100 nm.

In theory, the *noise figure* NF of a QD SOA should be better than the noise figure of a bulk SOA. Eq. (4.2) actually shows that the NF of an SOA basically depends on the inversion factor n_{sp} and the gain G,

$$\mathrm{NF} = 10\log_{10}\left(\frac{1}{G} + 2n_{sp}\frac{G-1}{G}\right). \qquad (4.2)$$

While the gain of a bulk and a QD SOA could be made similar, the (population) inversion factor of a QD SOA is close to ideal ($n_{sp} \approx 1$). This is due to the carrier reservoir (wetting layer) in QD SOA which allows an efficient filling of the QD states even for low current densities compared to bulk SOA. In practice, however, a typical FtF noise figure of a QD SOA is in the order of 5 dB [50], a number which is comparable to the FtF noise figure of the best bulk and QW SOA [21], see also *Chapter 2, Section 2.2.3*.

The *saturation input power* P_{sat}^{in} (see (4.3)) of an SOA depends on the modal cross-section C / Γ, the differential gain a and the effective carrier lifetime τ_c if only inter-band effects are considered,

$$P_{sat}^{in} = \frac{2hf_s \ln 2}{G_0 - 2} \frac{C}{\Gamma} \frac{1}{a} \frac{1}{\tau_c}. \qquad (4.3)$$

- Large values of P_{sat}^{in} are achievable with a large modal cross-section. QD SOA intrinsically offer large C / Γ of, i.e., $6 \times 10^{-13} \, m^2$, for example due to their low optical confinement factor. However, it needs to be mentioned that also broad-area structures with bulk active media are reported to have large values of P_{sat}^{in}, where a small Γ increases the saturation power as well [62].
- Low differential gain a also contributes to large saturation powers. In theory a is expected to increase if the dimensionality of the electronic system of the active medium reduces [31]. This is the case in QD active regions due to the delta-function like density of states which should lead to quite a strong change in the gain even for small carrier density variations. However, in practice the differential gain of QD SOA are not as large because the size of QD varies significantly, and the QD energy states are thus spread across a larger spectral range. It needs to be mentioned that the experimental results for the differential gain a for QW, bulk and QD SOA are comparable for the time being, and it is still a research topic to compare actual devices in terms of the differential gain.
- An effective carrier lifetime of $\tau_c \sim 100 \, ps \dots 1 \, ns$ is similar for QD, QW and bulk SOA. The effective carrier lifetime τ_c is related to the total effective carrier lifetime $\tilde{\tau}_c$ which influences the 90 %-10 % gain recovery time. Therefore, a speed enhancement of SOA devices is achieved with an increase in bias current or with a lifetime doping in bulk SOA and an associated reduction in τ_c. Further, longer SOA are preferable since the higher number of photons in the active medium causes the total effective carrier lifetime $\tilde{\tau}_c$ to decrease [131].
- It needs to be mentioned that the characteristic time in the saturation power relation changes for a QD SOA if ultra-fast effects are considered and a full population inversion (high number of carriers in the reservoir states) is assumed, see *Section 2.3.2*. Then, τ_c has to be substituted by the carrier capture time $\tau_1 \sim 1 \dots 10 \, ps$ from the wetting layer (reservoir states) into the QD states, and $\tau_1 \approx \tau_{SHB} \sim \varepsilon_{SHB}$ has to be associated with the SHB relaxation time. The significant reduction of this characteristic time causes QD SOA saturation powers to be larger by factors $30 \dots 100$ (continuous wave applications) or $5 \dots 10$ (50 ps pulse width applications) than for bulk SOA [111].

The *phase change* at the output of an SOA $\Delta\phi$ depends on the confinement factor Γ, the net modal gain change Δg, the device length L and the SOA *alpha factor* α_H, see (2.20). Typically, in bulk or QW SOA structures, a high confinement factor and an alpha factor α_H of value 2...8 are observed. Contrary to that, QD SOA with a population inversion of $n_{sp} = 1$ have an alpha factor $\alpha_H \approx 0$ mostly due to SHB, so that gain change and phase change are decoupled. Here, no or only small phase changes are imposed on the signal. This makes the QD SOA a promising device for linear amplification, i.e., for amplification without phase or amplitude distortion.

Recent publications of novel columnar QD SOA structures report also the possibility to observe a considerably high alpha factor. Such devices have been successfully used to achieve high-speed wavelength conversion and regeneration [132].

It can be concluded that in general the use of QD SOA is advantageous for linear applications. However, it needs to be mentioned that a well engineered bulk or QW SOA could be comparable or even better than its QD SOA counterpart.

4.2 Linear Amplifications Range

The capability of an SOA to perform linear amplification is limited to input signals that do not exceed a lowest and a highest input power level, independently of the modulation format, the symbol rate or the number of wavelength channels.

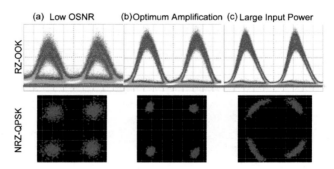

Fig. 4.1: Eye diagrams and constellations showing the limitations for linear amplification of data signals. (a)-(c) show the eye diagrams for a 10 Gbit/s RZ OOK signal as well as the constellation diagrams of a NRZ QPSK (4QAM) signal. In (a) the input signal into the SOA is very low resulting in a bad OSNR level at the output of the amplifier. In (b) error-free amplification of the data signal is observed, whereas in (c) for high input powers patterning effects due to compression of the gain occur and distort the OOK signals. In QPSK (4QAM) signals the phase change induced by a refractive index change within the SOA causes a nonlinear rotation of the constellation points, [B1].

For low input signal powers "error-free" amplification becomes an issue. In this work, the term "error-free" is used for distinct cases. The first case requires a signal quality which corresponds to a bit-error ratio (BER) of 10^{-9}. In the second case, the use of an advanced forward error correction (FEC) is assumed, which allows error-free operation for a raw BER of 10^{-3} or 10^{-5}, respectively. The mentioned limitations at low input powers are due to amplified spontaneous emission (ASE) noise which is almost independent of the signal input power. So if the input power decreases while the ASE remains constant a poor optical-signal-to-noise ratio (OSNR) will result for input signal powers that decrease below a certain value. An example of such OSNR limitation is presented in Fig. 4.1(a) for OOK and QPSK (4QAM) modulation formats. The eye diagram of the 10 Gbit/s return-to-zero (RZ) OOK single channel signal shows strong noise on the one and zero level indicating degraded eye quality. Further, the constellation diagram of the 20 GBd (40 Gbit/s) non-return-to-zero (NRZ) QPSK single channel signal shows a symmetrical broadening of the constellation points which causes a decrease of the signal quality.

For large input powers limitations occur as well. In this case the strong input powers will induce gain suppression (output power saturation) which leads to amplitude patterning and induces phase changes between subsequent bits. For example, the OOK format, see Fig. 4.1(c), shows bit-pattern effects. Such patterning is manifested by overshoots in the eye diagram and could lead to signal quality degradations. Patterning is seen when subsequent ones experience different gain values which happens when the recovery time between bits is comparable or exceeds the bit-slot duration. Since typical SOA recovery times are in the order of a few tens to hundreds of picoseconds, signal degradations are already visible at bit rates of a few Gbit/s.

At highest input powers even PSK or QAM encoded signals can degrade. For large input powers the SOA gain is suppressed. Transitions between symbols are affected by the complex SOA response. I.e., depending on the modulation format both the amplitude and phase fidelity of the amplification process is impaired to a different degree. Among the many implementations of M-PSK and M-ary QAM formats the best performing transmitters often use zero-crossing field strength transitions [133], and therefore generate power transitions ($-$), see Fig. 4.2(a,b). These power transitions change the carrier concentration N and therefore the SOA FtF gain $G_{ff} = \exp(g_{ff} L)$, where the FtF net modal gain g_{ff} is assumed to be independent of the SOA length L and comprises the SOA net modal chip gain g ($G = \exp(g L)$). A change $\Delta g_{ff} = \Delta(\ln G_{ff}) / L$ of the FtF net modal gain leads to a change Δn_{eff} of the effective refractive index n_{eff} by amplitude-phase coupling, which is described by the alpha factor α_H. With the vacuum wave number $k_0 = 2\pi / \lambda_s$ and the signal wavelength λ_s, the output phase change $\Delta\phi$ and effective refractive index change Δn_{eff} are related by $\Delta\phi = -k_0 \Delta n_{eff} L$. For the alpha factor we then find (see also (2.20), not regarding any input phase modulation)

$$\alpha_H = -2k_0 \frac{\partial n_{eff} / \partial N}{\partial g_{ff} / \partial N} \approx -2k_0 \frac{\Delta n_{eff}}{\Delta g_{ff}} = \frac{2\Delta\phi}{\Delta g_{ff} L} = \frac{2\Delta\phi}{\Delta(\ln G_{ff})} = \frac{2\Delta\phi}{\Delta(\ln G)}. \tag{4.4}$$

Thus, by amplitude-phase coupling in the SOA, gain changes induce unwanted phase deviations. An illustration of this effect is schematically depicted in Fig. 4.2 assuming a BPSK format and a saturated SOA. The transition between the two constellation points in Fig. 4.2(a) either involves a zero-crossing of the field (solid line) according to Fig. 4.2(b) with associated gain and phase changes of the SOA (Fig. 4.2(c,d)), or it maintains a constant envelope (dashed line) so that gain and phase changes do not happen, dashed lines in Fig. 4.2(b)-(d).

Going into more detail, the following can be seen. If the signal power reduces at time t_0, Fig. 4.2(b), the gain starts recovering from its operating point described by a saturated chip gain G_{Op} (given by the average input power) towards the unsaturated small-signal chip gain G_0. After traversing the constellation zero the signal power increases and reduces the gain towards its saturated value G_{Op}, Fig. 4.2(c). Coupled by the alpha factor, a gain change induces a refractive index change and therefore an SOA-induced phase shift $\Delta\phi$. SOA with a lower alpha factor ($\alpha_{H,1} < \alpha_{H,2}$) have less amplitude-to-phase conversion and therefore give rise to less phase changes, see Fig. 4.2(d), and also Fig. 4.1(c). As a consequence, SOA with lower alpha factors are expected to show better signal qualities for phase-encoded data with a high probability of large power transitions. The constellation diagrams with and without induced phase errors are presented in Fig. 4.2(e,f).

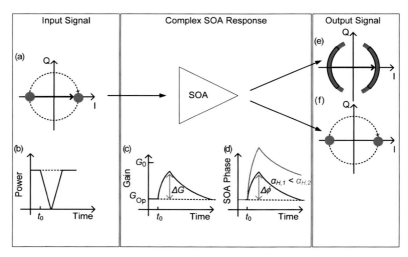

Fig. 4.2: Response of a saturated SOA in reaction to a BPSK (2QAM) signal with two possible transitions from symbol to symbol. (a) BPSK constellation diagram with in-phase (I) and quadrature component (Q) of the electric field. Solid line: zero-crossing transition; dashed line: constant-envelope transition. (b) Time dependencies for the two types of power transitions. SOA response that affects the (c) gain and (d) refractive index which leads to an SOA-induced phase deviation $\Delta\phi$. SOA with lower alpha factors induce less amplitude-to-phase conversion and therefore amplify the electric input field with a better phase fidelity. BPSK constellation diagram after amplification with a saturated SOA for (e) zero-crossing transition (for two alpha factors) and (f) constant-envelope transition, [J1].

Typically, the phase recovery in SOA is slower than the gain recovery [112], [134], so a phase change induced at the power transition time has not necessarily died out at the time of signal decision (usually in the center of symbol time slot), so that the data phase is perturbed and errors occur. In addition to phase errors amplitude errors are expected to occur in *M*-ary QAM signals (e.g., in 16QAM). Because *M*-ary QAM signals comprise multiple symbols with different amplitude levels extreme transitions from one corner to the other are less likely.

Thus, phase errors due to amplitude-phase coupling are less likely as well. In average the amplitude distance between symbols reduces due to gain saturation.

Some examples of amplitude transitions in constellation diagrams for practical PSK and *M*-ary QAM implementations are shown in Fig. 4.3(a) for BPSK, QPSK and 16QAM. The transition probabilities for all occurring normalized amplitude changes are depicted in Fig. 4.3(b). The transition probability of the largest amplitude change reduces from 50 % for BPSK to 25 % at QPSK down to below 5 % for 16QAM. Thus, the probability to observe a large amplitude change decreases the higher the order of the modulation format.

If more wavelength channels are involved, cross-gain modulation (XGM) and cross-phase-modulation (XPM) can cause signal quality degradations by inter-channel cross-talk. Such distortions are only relevant if the total average input power of the channels exceeds the saturation input power and thus the SOA is operated in the nonlinear gain regime, Fig. 2.11.

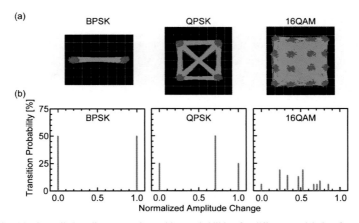

Fig. 4.3: Constellation diagrams and transition probabilities for different modulation formats. (a) Constellation diagrams with amplitude and phase transitions for BPSK (2QAM), QPSK (4QAM) and 16QAM data signals. (b) Transition probability as a function of the amplitude change normalized to the largest possible amplitude transition. The probability of large transitions decreases for higher order modulation formats, [J1].

4.3 Input Power Dynamic Range for OOK Data Signals

Next-generation (NG) long-reach PON [135]-[137] will allow the consolidation of thousands of CO and can be implemented in a hybrid time-division-multiplexing (TDM)/WDM approach with direct detection. One option to increase the tolerable loss budget is the use of optical amplifiers [138], [139]. Here, the burst mode nature of the upstream path data together with the probable existence of a variety of different split ratios require amplifiers with highest IPDR. In the past, various optical amplifier technologies have been discussed, which only partly satisfied this requirement. Ordinary fiber-doped amplifiers showed burst-sensitive operation. Thus, this amplifier technology needs gain compensation techniques [138], [139]. A hybrid technology, using SOA together with Raman amplifiers, proved to provide large bandwidths [140]. Yet, this hybrid solution still requires a Raman pump laser and offers moderate IPDR. On the contrary, SOA are a promising and cost efficient technology [141] which is studied in the following.

In this section, the IPDR of SOA are discussed. Guidelines to design SOA with large IPDR are given. For this purpose, an IPDR description is introduced and the respective parameters are determined. It is shown, that QD SOA with a low number of QD-layers (low optical confinement) and long devices provide potentially the best IPDR for a particular gain. In single and two-channel experiments the QD SOA is operated error-free at bit rates of 2.5 Gbit/s, 10 Gbit/s and 40 Gbit/s with OOK modulation. IPDR dependance on wavelength and bias current is addressed.

4.3.1 Description of Input Power Dynamic Range

For a mathematical introduction of the IPDR, a signal quality measure needs to be introduced. The benchmark of a digital communication system is performed in terms of the BER, also called bit-error probability. The BER relates the number of received bit errors to the number of transmitted bits. Most often a direct measurement of the BER is complex and time consuming, e.g., if a low bit rate signal is investigated several minutes or even tens of minutes are required to measure a statistically relevant BER in the range of 10^{-9}. Therefore the quality factor Q is introduced which is a powerful tool to estimate the BER out of a received eye diagram.

The Q factor is defined with the expectation of voltage for a received one u_1, the expectation of voltage for a received zero u_0 and the standard deviations of the one level σ_1 and the zero level σ_0, respectively

$$Q = \frac{u_1 - u_0}{\sigma_1 + \sigma_0}. \tag{4.5}$$

Often, the Q factor is expressed in linear units, or as a Q^2 factor in logarithmic units,

$$Q^2\big|_{dB} = 10\log_{10} Q^2 = 20\log_{10} Q. \tag{4.6}$$

The Q factor can be related to a minimum BER if the following assumptions are fulfilled [61], [91], [115]:

- The probability density functions of the sampled input voltage of the decision circuit for the received symbols zero and one are described by Gaussian distributions.
- Any interference of a received impulse with impulses at neighboring sampling points disappears (no intersymbol interference).
- Optimum decision threshold is used.
- Standard deviations of the one level and the zero level are approximately identical.

• The product of the probability of receiving a bit one and the standard deviation of the zero level is equal to the product of the probability of receiving a bit zero and the standard deviation of the one level.

$$BER = \frac{1}{2} \text{erfc}\left(\frac{Q}{\sqrt{2}}\right). \tag{4.7}$$

Ordinarily, the computation of the BER is very difficult due to the fact that the exact shape of the probability density functions for the decision circuit voltages at the sampling times are unknown. Therefore, no unique dependency between the signal-to-noise ratio (SNR) and the BER can be established. However, because the Gaussian distribution is fully determined by moments up to the second order, and if the noise signals at the decision circuit input is truly Gaussian, then the BER and the SNR can be related. If further usual impulse shapes (raised cosine) are received, the bit-error parameter Q and the BER may be deduced from a measurement of the SNR

$$SNR = (0.97...1) \cdot Q^2 \rightarrow SNR = Q^2. \tag{4.8}$$

In the range of 9 dB $< Q^2 < 18$ dB, the calculated BER usually corresponds nicely to a directly measured BER. However, for values < 9 dB and values > 18 dB, the Q^2 factor method leads to large errors between a calculated BER and a directly measured BER, because even small variations in the Q^2 factor cause large variations in the BER.

As long as the communication system is noise limited, a Gaussian approximation of the noise statistics typically holds, and the results are reliable. If nonlinear distortions and inter-symbol interference occur, e.g., due to SOA patterning effects, the exact shape of the probability density functions can be significantly altered. For example, transient overshoots typically lead to an underestimated quality factor and thus an overestimated BER [142]. However, the Q^2 factor is measured using a 20 % or 10 % window in the middle of the eye diagram. Thus, if the bit-slot histogram is generated the deviations of the real probability density function from a Gaussian distribution can be assumed to be rather small in most cases. In this work, the Gaussian distribution is used as an approximation, and it has been verified by direct BER measurements that the deviations are small. In recent years, the Q^2 factor definition is also applied to differential phase-shift keying (DPSK) or differential quadrature phase-shift keying (DQPSK) data signals [143], see *Appendix A.6*.

The IPDR is defined as the SOA input power range within which error-free amplification is achieved, see Fig. 4.4(a). In this section related to the IPDR of OOK signals, the term error-free is used for two cases. The first case requires a signal quality of $Q^2 = 15.6$ dB which corresponds to a BER of 10^{-9}. In the second case an advanced FEC allows error-free operation for Q^2 of 12.6 dB corresponding to a BER of 10^{-5}. A Q^2 of 12.6 dB is better than what will be required for NG-PON where bit-error ratios around 10^{-4} are tolerated together with FEC [12].

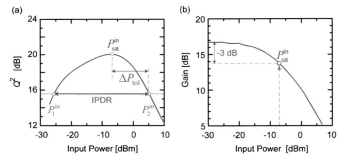

Fig. 4.4: Definition of the input power dynamic range (IPDR). (a) The IPDR is defined as the SOA input power range within which error-free amplification is achieved. The lower and upper limits of the IPDR are due to a low OSNR and gain saturation, respectively. The plot also visualizes the definition of the empirical tolerance factor $\Delta P_{tol} = 10 \lg p_{tol} = 10 \lg(P_2^{in} / P_{sat}^{in})$. (b) Gain as a function of input power. The 3 dB saturation input power is marked, [J2].

The IPDR limitation P_1^{in} on the low input power side is due to a low OSNR whereas the IPDR limitation P_2^{in} on the high input power side is due to gain saturation induced patterning and inter-channel cross-talk (multi-channel experiments). The IPDR is defined as

$$\text{IPDR} = 10 \lg(P_2^{in}) - 10 \lg(P_1^{in}). \tag{4.9}$$

In the following, an IPDR description is introduced to derive device design guidelines.

The lower limit of the IPDR P_1^{in} may be derived starting with the well known electrical SNR equation for incoherent direct detection with optical amplification [92] considering a reasonable unsaturated gain $G_0 \gg 1$, (2.29). Here, only the dominant signal-noise beating (first term in denominator) and noise-noise beating (second term in denominator) terms are considered. $P_{ASE\parallel}$ is the amplified spontaneous emission power from the SOA, which is co-polarized with the signal, P^{in} is the optical input power to the SOA, B_O the optical filter bandwidth and B_e the electrical filter bandwidth. Resolving (2.29) for the lower limit of the IPDR P_1^{in} for the case of error-free amplification $\text{SNR}_{min} = Q_{min}^2$ gives

$$P_1^{in} = \left[\sqrt{2\text{SNR}_{min} B_e B_O + 4\text{SNR}_{min}^2 B_e^2} + 2\text{SNR}_{min} B_e \right] P_{ASE\parallel}/G_0 B_O \quad , \tag{4.10}$$

with

$$P_{ASE\parallel} = n_{sp}(G_0 - 1) \, hf_s \cdot B_O. \tag{4.11}$$

Here, n_{sp} is the inversion factor, h is the Planck constant and f_s is the optical carrier frequency.

The upper limit of the IPDR P_2^{in} is limited by bit patterning induced by gain saturation. We derive P_2^{in} with a rate-equation SOA model neglecting nonlinearities, facet reflectivity and ASE noise for time-independent quantities [62]. Further, a linear relation between the modal net gain g_0 and the carrier density N_0 is assumed. Then, we write

$$P_2^{in} = p_{tol} P_{sat}^{in}, \qquad P_{sat}^{in} = \frac{hf_s}{G_0 - 2} \frac{2\ln(2)}{\Gamma} \frac{C}{a} \frac{1}{\tau_c} \quad , \tag{4.12}$$

with

$$G_0 = \exp(gL) = \exp(a\,(N_0 - N_t)\,L), \qquad N_0 = r\tau_c. \tag{4.13}$$

p_{tol} describes an empirical tolerance factor of which the upper IPDR limit P_2^{in} is larger than the steady-state 3 dB saturation input power P_{sat}^{in}, Fig. 4.4(b). In Fig. 4.4(a) the tolerance factor is expressed in logarithmic units $\Delta P_{tol} = 10\lg p_{tol} = 10\lg P_2^{in} - 10\lg P_{sat}^{in}$. C is the area of the active region, Γ the optical confinement factor, a is the differential gain, τ_c is the effective carrier lifetime, L is the length of the device, N_t is the carrier density at transparency and $r = I/eV$ is the carrier generation rate with the bias current I, the active volume V and the elementary electric charge e. It should be mentioned that the carrier density N_0 is understood as the number of all carriers in the device [111].

4.3.2 1.3 µm SOA Device Structure and Characterization

The SOA used in this study are the devices SOA 1 and SOA 2 introduced in the chapter 3. The applied current is 490 mA for SOA 1 (QD SOA) in all measurements and varied in the measurements with SOA 2 (QD SOA).

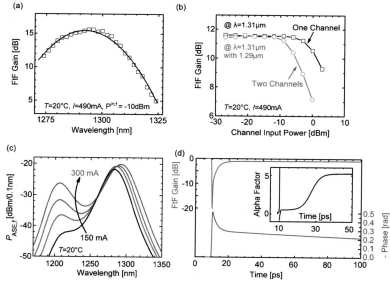

Fig. 4.5: Typical 1.3 µm QD SOA characteristics of the measured devices. (a) A 3 dB gain bandwidth of 35 nm at 1.295 µm with a peak gain of 15.5 dB is found with SOA 1. (b) FtF gain as a function of the input power, showing linear behavior over a huge range of input power levels. The gain is constant within a 1 dB margin for channel input power levels of up to 0 dBm for the single-channel case and up to –6 dBm for the two-channel case. (c) Typical ASE power spectrum for different bias currents at SOA 2. (d) The gain and the phase response of the QD SOA are plotted. A fast (1…10 ps) and a slow (100…200 ps) time constant are visible. The overall phase effects are weak though and dominated by the slow component. Inset shows low time-dependent effective alpha factor measured on SOA 2, [J2].

A typical FtF gain spectrum of these devices is shown in Fig. 4.5(a), measured at SOA 1. The gain peak is around 1295 nm. For this steady-state characterization the temperature is set to 20° C and the input power is −10 dBm. Fig. 4.5(b) shows the linear behavior of this device for a large range of input power levels. The gain G_{ff} is constant within a 1 dB margin for channel input powers between −30 dBm and 0 dBm for one cw channel at 1310 nm and up to −6 dBm for two channels measured on 1310 nm with another cw signal launched at 1290 nm. Such degree of linearity is only known from the linear optical amplifier approach [125], specifically designed SOA using strained bulk or multi-quantum well structures [22], [144], [145] and holding beam techniques in SOA [146]. Fig. 4.5(c) shows the ASE power spectrum of SOA 2 for different bias currents measured with a resolution bandwidth of 0.1 nm. The QD ground state gain increases with current (gain peak around 1.3 μm). The maximum ground state gain is reached at a bias current of around 300 mA due to complete inversion. A further bias current increase leads to a reduction in ground state gain due to device heating. The excited state gain peak is close to 1.2 μm. The excited states are not fully inverted within the range of currents measured here.

The 3 dB in-fiber saturation input powers $P_{sat}^{in,f}$ and the unsaturated FtF gain G_{f0} of SOA 1 and SOA 2 are presented in Table 4.1 for two wavelengths. The bias current of SOA 1 and SOA 2 is 490 mA and 300 mA, respectively. The saturation input power levels of the two samples are not comparable due to different unsaturated gains, waveguide widths and supplied current densities which are around 6 kA/cm² for SOA 1 and around 2 kA/cm² for SOA 2.

	Wavelength 1290 nm		Wavelength 1310 nm	
	SOA 1	SOA 2	SOA 1	SOA 2
Saturation input power $P_{sat}^{in,f}$	2 dBm	−7 dBm	4 dBm	−4 dBm
Unsaturated gain G_{f0}	15.5 dB	17 dB	11.5 dB	14 dB

Table 4.1: In-fiber saturation input power and unsaturated FtF gain as a function of wavelength for SOA 1 and SOA 2. SOA 1 is operated with a bias current of 490 mA and SOA 2 is operated with 300 mA. The bias currents correspond to current densities of around 6 kA/cm² for SOA 1 and 2 kA/cm² for SOA 2, respectively. The temperature is set to 20 °C.

The dynamic behavior of the QD SOA in a pump-probe experiment is shown in Fig. 4.5(d). The QD SOA (SOA 2) has a fast and a slow carrier refilling time as discussed in *Section 2.3* and *Chapter 3*. The inset in Fig. 4.5(d) shows that the effective alpha factor is not constant. It strongly changes with time under a large signal modulation [112] as discussed in *Section 2.2.6*.

4.3.3 Input Power Dynamic Range Dependance on Bit rate

The IPDR of the 1.3 μm QD SOA is studied for single and multiple data signals at various bit rates. Experiments are performed with the setup shown in Fig. 4.6. Two decorrelated data signals (generation see _Appendix A.1_) are adjusted to various power levels before launching the composite signal into the SOA. After the SOA one channel is blocked by a tunable filter, while the Q^2 factor of the remaining data channel is analyzed with a digital communications analyzer (DCA, reception see _Appendix A.3_). The receiver is operated 10 dB above its sensitivity threshold of $Q^2 = 15.6$ dB.

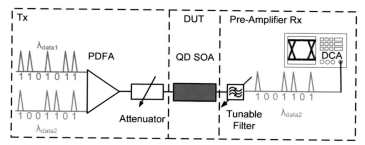

Fig. 4.6: Setup to measure IPDR of the SOA device under test (DUT). The quality of the OOK data signals (Q^2 factor) after amplification in a SOA are measured for single and two channel operation at bit rates of 2.5 Gbit/s, 10 Gbit/s, and 40 Gbit/s while the input power is varied. In the two channel operation mode the signals are decorrelated. The quality of the received signal is measured with a pre-amplifier receiver, [J2].

We investigate the IPDR for bit rates of 2.5 Gbit/s NRZ OOK, 10 Gbit/s NRZ OOK, and 40 Gbit/s RZ OOK with SOA 1. Firstly, a single data signal at a wavelength of 1310 nm is generated. Its power at the SOA input (measured in fiber) is varied, and the quality of the received signal is measured. Secondly, two decorrelated data signals at a wavelength of 1310 nm and 1290 nm are generated. Their power is kept at the same level and varied from −27 dBm to 5 dBm.

The sensitivity of the Q^2 factor to variations in the power launched into the SOA for the single channel case is shown in Fig. 4.7(a). An IPDR exceeding 25 dB is found at all bit rates for a Q^2 factor of 15.6 dB (grey line within the figure). The IPDR for a Q^2 factor of 12.6 dB is estimated to exceed 32 dB for all bit rates (light blue line within the figure).

The small deviations in Q^2 factor between the bit rates at low launch powers arise from electrical bandwidth adjustments for different bit rates at the DCA and the use of a precision time base at 40 Gbit/s only.

In the high input power limit, the signal quality is affected by patterning that dominates when the SOA is operated in saturation and the QD SOA dynamics are too slow to follow the pattern [147]. At 2.5 Gbit/s both the slow gain dynamic of the reservoir as well as the fast QD refilling are clearly able to follow the data-stream. We therefore find error-free operation up to highest input powers. At 10 Gbit/s where the QD SOA is operated with a bit-period that is comparable to the slow 100 ps carrier refilling time we observe strong overshooting, see eye diagram in Fig. 4.7. At 40 Gbit/s the influence of the slow carrier refilling leads to patterning. Since the pulses are shorter than the slow carrier refilling time the effect statistically averages out. The patterning is therefore not too severe. Conversely the influence of the fast QD gain dynamic on the amplification process increases. Due to the fact that the fast effect has the same time scale as the bit-period of the 40 Gbit/s data signal, it leads to a gain compression that acts almost equally onto all bits. This explains why the IPDR of the 40 Gbit/s RZ data signal is comparable to the 10 Gbit/s NRZ data signal. Fig. 4.7 shows also the eye diagrams at the lower and upper limit of the IPDR for the different bit rates.

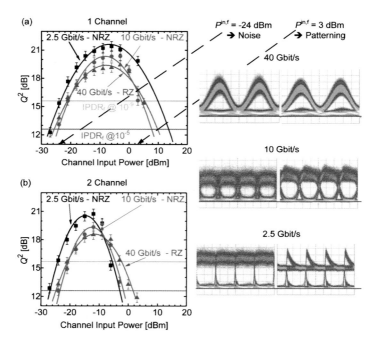

Fig. 4.7: Q^2 factor of one and two channels amplified with a QD SOA versus channel input power at bit rates of 2.5, 10 and 40 Gbit/s. (a) In the single channel case we find an IPDR$_f$ = 10lg ($P_2^{in,f}$ / $P_1^{in,f}$) (Q^2 = 12.6 dB marked by the light blue line) between 32 dB and 41 dB for 2.5 Gbit/s, 10 Gbit/s and 40 Gbit/s, respectively. Eye diagrams show the IPDR limitations in the single channel case. For high input powers the signal quality is limited by overshooting (2.5 Gbit/s) and patterning (10 Gbit/s and 40 Gbit/s). For low input powers the signal quality is limited by noise. (b) In the two channel case we find an IPDR$_f$ (Q^2 = 12.6 dB) in the range of 22 dB to 27 dB for bit rates of 2.5, 10 and 40 Gbit/s. The IPDR$_f$ results for a Q^2 = 15.6 dB are marked by the grey line, [J2].

The results of the two channel experiments are plotted in Fig. 4.7(b). The IPDR becomes almost independent of the bit rate.

At low input powers, the signal qualities are similar to the single-channel case. This is to be expected since the SOA is unsaturated and operation is only noise limited.

In the high input power limit, the SOA is operated in saturation and we observe signal degradations due to patterning from one channel and also due to inter-channel cross-talk [148]. The upper IPDR limit is reached when the total in-fiber input power approaches the in-fiber saturation input power. For the SOA 1 with a 3 dB in-fiber saturation input power of 4 dBm at a wavelength of 1310 nm (see Table 4.1) the upper IPDR limit therefore is close to 0 dBm for each of the two channels. This behavior may be understood by the fact that the inter-channel cross-talk between two decorrelated channels may occur with a random time delay. As a consequence the cross-talk between the channels leads to a degradation which is almost independent of the tested bit rates. The absolute values of the $IPDR_f = 10lg\ (P_2^{in,f} / P_1^{in,f})$ measured with SOA 1 at 1310 nm for the two decorrelated signals at, e.g., 10 Gbit/s are 16 dB and 22 dB at Q^2 of 15.6 dB and 12.6 dB, respectively. These values, measured with the QD SOA, are comparable with the results reported for the linear optical amplifier (LOA) in [125].

4.3.4 Input Power Dynamic Range Dependance on Wavelength

The results for the IPDR depend on the wavelength of the data signal. This is mostly due to the wavelength dependance of the unsaturated gain G_0 which influences the saturation input power, see Eq. (4.12). The IPDR results for measurements at wavelengths of 1290 nm and 1310 nm with SOA 1 are summarized in Table 4.2. Measurements are performed at bit rates of 2.5 Gbit/s, 10 Gbit/s and 40 Gbit/s for the single and also for the two channel case. For the two channel experiments, the decorrelated data signals always carry the same bit rate and the same channel input power.

The spectral IPDR dependance to the most part correlates with the spectral gain dependance. The higher the gain and thus the lower the saturation input power the lower the IPDR in the single channel case. This agrees well with Eq. (4.12) and is in agreement with the results in Table 4.2, where the $IPDR_f$ at 1310 nm (11.5 dB FtF gain, in-fiber saturation input power of 4 dBm) exceeds the $IPDR_f$ found at 1290 nm (15.5 dB FtF gain, in-fiber saturation input power of 2 dBm). The $IPDR_f$ at 1310 nm exceeds the $IPDR_f$ at 1290 nm by 5 dB at 2.5 Gbit/s, by 7 dB at 10 Gbit/s and by 2 dB at 40 Gbit/s.

Bit rate [Gbit/s]	$IPDR_f$ [dB] 1 channel measured at				$IPDR_f$ [dB] 2 channels measured at			
	1290 nm		1310 nm		1290 nm		1310 nm	
2.5 – NRZ OOK	36	28	41	33	28	23	24	19
10 – NRZ OOK	25	18	32	25	24	18	22	16
40 – RZ OOK	33	24	36	26	28	20	27	19
BER (Q^2 measured)	10^{-5}	10^{-9}	10^{-5}	10^{-9}	10^{-5}	10^{-9}	10^{-5}	10^{-9}

Table 4.2: Summary of $IPDR_f$ dependance on wavelength, bit rates, signal quality factor (BER) and channel number. In the single channel case, the IPDR for different bit rates is larger at wavelength with lower gain. In the two channel case, the IPDR measured at the gain peak is larger due to reduced cross-gain modulation distortions.

In the 2 channel situation, the spectral dependance of the IPDR is affected most by the channel at the gain peak wavelength. I.e., the channel around the gain peak (1290 nm) degrades the channel measured at 1310 nm (outside of the gain peak) to larger extent than vice versa. More precisely, the $IPDR_f$ measured at 1290 nm exceeds the $IPDR_f$ at 1310 nm at all bit rates. However, the signal degradations still stem from patterning from one channel and from inter-channel cross-talk. The $IPDR_f$ differences between 1290 nm and 1310 nm are 4 dB at 2.5 Gbit/s, 2 dB at 10 Gbit/s and 1 dB at 40 Gbit/s.

4.3.5 Input Power Dynamic Range Dependance on Rival Signal

Results of a SOA IPDR worst case scenario, i.e., with a rival signal at the gain peak, are presented in Fig. 4.8 for experiments performed with SOA 2. A 10 Gbit/s NRZ OOK data signal at a wavelength of 1290 nm is launched into the QD SOA together with a 40 Gbit/s RZ OOK data signal at a wavelength of 1310 nm. The first serves as a rival signal on the 40 Gbit/s signal. A 10 Gbit/s signal was chosen as its bit-slot period is in the order of the slow gain dynamic of the device caused by the reservoir refilling. This represents a case with large cross-gain and cross-phase modulation distortions. The $IPDR_f$ of the 40 Gbit/s signal is analyzed for three different input power levels of the rival signal.

First, a low in-fiber input power in the unsaturated gain regime of the SOA of -12 dBm is chosen. Then, a moderate in-fiber input power of -9 dBm and finally an input power of -6 dBm is used which is saturating the gain of SOA 2 (see Table 4.1). Despite the rival signal, the $IPDR_f$ is large and exceeds 25 dB for a Q^2 of 15.6 dB as long as the input power level of the rival signal is moderate (< -9 dBm). Higher input power levels of the rival signal cause maximum inter-channel cross-talk which decreases the IPDR of the 40 Gbit/s signal. SOA 2 is operated with a bias current of 300 mA at a temperature of 20 °C.

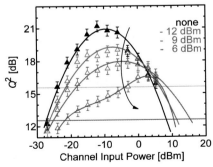

Fig. 4.8: $IPDR_f$ dependance on input power for a rival signal at the gain peak. A 40 Gbit/s RZ OOK data signal at a wavelength of 1310 nm is launched to the QD SOA together with a rival 10 Gbit/s NRZ OOK data signal at a wavelength of 1290 nm. The IPDR of the 40 Gbit/s signal is analyzed for three different input power levels of the rival signal. The IPDR is large as long as the input power level of the distorting signal is moderate. The grey line indicates the $IPDR_f$ results for a Q^2 of 15.6 dB whereas the light blue line indicates the $IPDR_f$ results for a Q^2 of 12.6 dB, [J2].

4.3.6 Input Power Dynamic Range Dependance on Bias Current

The IPDR dependance on bias current at a given gain is investigated for a single channel experiment with a 40 Gbit/s RZ OOK data signal. When adapting the bias currents from 150 mA to 350 mA the wavelength of the test signal is tuned such that the unsaturated FtF gain G_{f0} of SOA 2 maintains 15 dB. This tuning is performed to eliminate effects from the gain onto the saturation power. The wavelength is thus changed from 1289.5 nm ($I = 150$ mA) to 1307 nm ($I = 350$ mA).

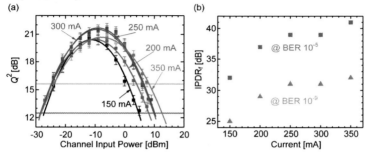

Fig. 4.9: IPDR$_f$ dependance on bias currents for SOA 2. (a) Signal quality measurement over input power for different bias currents for a 40 Gbit/s RZ OOK signal. (b) IPDR$_f$ derived from (a) as a function of the bias current. The higher the current is the larger the IPDR$_f$. It exceeds 30 dB for a Q^2 of 15.6 dB and 40 dB for a Q^2 of 12.6 dB. Throughout the measurements the wavelength has always been adjusted to keep a constant gain of 15 dB, [J2].

The signal quality as a function of channel input power for different bias currents is presented in Fig. 4.9(a). The IPDR$_f$ for a Q^2 of 12.6 dB as well as for Q^2 of 15.6 dB is indicated. Fig. 4.9(b) shows the IPDR$_f$ as a function of bias current for Q^2 of 12.6 dB (light blue squares) as well as for Q^2 of 15.6 dB (grey triangles). The maximum IPDR$_f$ is exceeding 40 dB for a Q^2 of 12.6 dB and 30 dB for a Q^2 of 15.6 dB. In general, the IPDR improves with larger bias currents. The IPDR$_f$ increases monotonously up to 41 dB for a Q^2 of 12.6 dB and up to 32 dB for a Q^2 of 15.6 dB. From Fig. 4.9(a) it can be seen that the lower input power limit of the IPDR$_f$ hardly changes with the bias current. However, the upper IPDR$_f$ limit increases monotonously with increasing bias current. It is attributed that this increase of the saturation input power is mostly due to the reduction of the effective carrier lifetime, see (4.12), since the unsaturated gain and the modal cross-section are constant and therefore they cannot contribute to the improvement. The effective carrier lifetime in (4.12) can be interpreted as the slow carrier refilling time of the reservoir states in the QD SOA. A larger bias current causes a higher population of the reservoir states and thus an efficient and fast refilling of the QD states, see *Section 2.3.2*.

4.3.7 SOA Design Guidelines for Large IPDR with Moderate Gain

In this subsection design guidelines to optimize the IPDR$_f$ of an SOA are given. For this the IPDR description derived in subsection 4.3.1 is used. The IPDR description is first validated with parameters derived from the experiments presented in subsections 4.3.2 and 4.3.6. Actually, all required parameters can be derived from steady-state characterization and 40 Gbit/s OOK measurements of the IPDR$_f$ at different currents and at a constant FtF gain of 15 dB as described in Fig. 4.5(c), Fig. 4.9(a) and Fig. 4.11(a).

It is shown, that low QD-layer numbers and long devices provide the best IPDR for a particular gain.

4.3.7.1 Experimental Parameter Determination for IPDR Description

To use the IPDR description for P_1^{in} and P_2^{in} as derived in (4.10) and (4.12) measured values of the saturation input power P_{sat}^{in}, the unsaturated gain G_0 and the ASE output power P_{ASE} are needed. For this a steady-state and dynamic measurements on SOA 2 at various currents and input powers are performed. The IPDR description of the SOA device was derived for chip-to-chip values (P_1^{in}, P_2^{in}, G_0), while in the experiment fiber-to-fiber values ($P_1^{in,f}$, $P_2^{in,f}$, G_{f0}) are measured. The measured and chip values are adapted by coupling losses of 4 dB per facet.

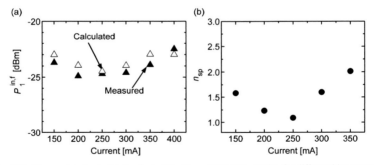

Fig. 4.10: Measured and calculated values of the lower limit of the IPDR$_f$ for different bias currents and the inversion factor. (a) Measured values of the lower limit of the IPDR$_f$ for different operating points (filled triangles) showing good agreement with the calculated values (open triangles). The calculated values are obtained applying the IPDR description to the ASE output power and the unsaturated gain from steady-state characterization. (b) The inversion factor as a function of current. The inversion factor reaches its lowest value of almost 1 at a current of 250 mA. The values are derived from the steady-state characterization of the ASE output power and the unsaturated gain, [J2].

To calculate the lower limit of the IPDR$_f$ $P_1^{in,f}$ according to (4.10) the values of the steady-state measurements of $P_{ASE,f}$ (ASE power already polarized, thus no index ||) from Fig. 4.5(c) at an unsaturated gain G_{f0} = 15 dB are used. As mentioned in Fig. 4.9 the wavelength is tuned to keep G_{f0} at a value of 15 dB. Here the system parameters of the optical bandwidth B_O and the electrical bandwidth B_e of a 40 Gbit/s system are used, see Table 4.3. The calculated values for the lower IPDR$_f$ limit (open triangles in Fig. 4.10(a)) are compared with the IPDR$_f$

measurement results from Fig. 4.9(a) for a 40 Gbit/s signal at a Q^2 of 15.6 dB (filled triangles in Fig. 4.10(a)). It can be seen that calculated and measured values agree well.

The lowest P_1^{in} is obtained when the inversion factor is smallest which is to be expected from theory, (4.10) and (4.11). This behavior is supported by the plots shown in Fig. 4.10(a) and (b). Fig. 4.10(a) shows the $P_1^{in,f}$ versus the SOA bias currents. The lowest value is found at $I = 250$ mA. Fig. 4.10(b) shows the calculated inversion factor. Its value is lowest at $I = 250$ mA, which was to be expected from the IPDR description. The inversion factor n_{sp} in Fig. 4.10(b) has been calculated with (4.11) using the experimental data for $P_{ASE,f}$ at the gain $G_{f0} = 15$ dB taking into account the respective signal wavelengths (corresponding photon energy hf_s) at a specific bias current. n_{sp} decreases for increasing current levels up to 250 mA. This is due to a higher population inversion of the QD states. A further current increase leads to carrier excitations from the QD ground state due to thermal issues which causes an increase in the inversion factor.

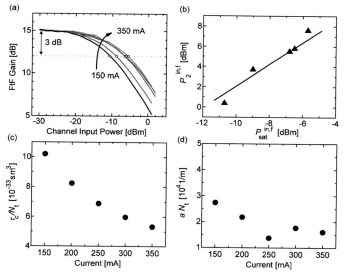

Fig. 4.11: Important device parameters of SOA 2. (a) FtF gain as a function of input power for different bias currents. The 3 dB in-fiber saturation input powers at various bias currents are marked by circles. They increase with the bias currents. For the plots at different currents the wavelength of the data signal is always adjusted to guarantee a constant unsaturated FtF gain of 15 dB. (b) $P_2^{in,f}$ versus $P_{sat}^{in,f}$ for different bias currents. It can be seen that $P_2^{in,f}$ and $P_{sat}^{in,f}$ are always offset by 12 dB. The values for $P_2^{in,f}$ are derived from Fig. 4.9(a) and the values for $P_{sat}^{in,f}$ are obtained from Fig. 4.11(a). (c) The effective carrier lifetime per carrier density at transparency for different current values as derived from the unsaturated gain and the saturation input power. A higher carrier density due to higher current reduces the effective carrier lifetime. (d) Differential gain times the carrier density at transparency for different current values derived as in (c). The lowest differential gain is achieved at a current of 250 mA, [J2].

In the following, the upper limit of the IPDR$_f$ $P_2^{\text{in,f}}$ is discussed. Fig. 4.11(a) shows the FtF gain versus the input power for different bias currents for the discussed measurement procedure. The 3 dB in-fiber saturation input powers $P_{\text{sat}}^{\text{in,f}}$ are indicated by circles. $P_{\text{sat}}^{\text{in,f}}$ increases with the bias current. Next the upper limit of the IPDR$_f$ $P_2^{\text{in,f}}$ is extracted at different bias currents from the measured results in Fig. 4.9(a) and the in-fiber saturation input power $P_{\text{sat}}^{\text{in,f}}$ versus the upper limit of the IPDR$_f$ $P_2^{\text{in,f}}$ is plotted, see Fig. 4.11(b). It may be seen that the QD SOA upper IPDR$_f$ limit is offset from the steady-state in-fiber saturation input power by the tolerance factor ΔP_{tol} of around 12 dB. The effective carrier lifetime τ_c as well as the differential gain a cannot easily be measured. However, the device parameters τ_c / N_t and aN_t can be derived from measurements using (4.12) and (4.13). The results are plotted in Fig. 4.11(c) and (d) as a function of the bias current. It is obvious that the effective carrier lifetime (reservoir refilling time) reduces for operating points with higher bias currents due to a larger occupation of the reservoir. The differential gain only slightly changes.

Now, all parameters of the device and the system are known which enable us to predict design guidelines for large IPDR$_f$ with moderate gain. A summary of all the parameters is shown in Table 4.3.

4.3.7.2 Guidelines to Optimize SOA for Largest IPDR with Moderate Gain

The IPDR description provides guidelines for designing SOA with large IPDR$_f$ and moderate FtF gain. Important parameters that can be controlled are the device length L, the number of QD-layers l, the confinement factor Γ, the waveguide cross-section C, the doping concentration of the active region material and the bias current I. The following discussion is similar to the discussion performed in *Section 4.1*, but here the interrelation of the parameters are important in optimization of both the gain and the IPDR. The upper limit of the IPDR$_f$ $P_2^{\text{in,f}}$ is given by (4.12) and (4.13). For a given unsaturated FtF gain G_{f0} and a specific ΔP_{tol} we can write

$$P_2^{\text{in,f}} \propto \frac{1}{G_{f0}} \frac{C}{\Gamma} \frac{1}{\tau_c} \frac{1}{a}, \qquad G_{f0} = \alpha_{\text{coupling}} \left\{ \exp\left[a\left(\frac{I\,\tau_c}{eV} - N_t \right) L \right] \right\} \alpha_{\text{coupling}} . \qquad (4.14)$$

This leads us to the following strategy to maximize $P_2^{\text{in,f}}$:

- Choose G_{f0} as small as possible to serve the purpose: From (4.14), it can be seen that a minimum gain must be stipulated as the IPDR$_f$ could go to infinity for smallest gain values.
- Large modal cross-section C / Γ with small confinement factor. This is obtained by adapting the device geometry and choosing the SOA type accordingly, e.g., the confinement factor of a QW or QD SOA is lower than the confinement factor of a bulk SOA. However, a decrease in the confinement causes a decrease in the unsaturated gain which can be compensated by an increase in the device length and/or a higher bias current.

- Lifetime-doping of the active region material [106]. A lifetime-doping can cause a reduction of the effective carrier lifetime (not necessarily in all SOA). However, the gain is again decreased and the already discussed measures have to be taken.
- High bias current I. The increase in bias current causes a reduction in the effective carrier lifetime. Then, the unsaturated gain increases. It can be compensated by reducing the SOA length or the confinement. It needs to be mentioned that the unsaturated gain cannot be increased arbitrarily. Thermal effects might decrease the inversion.
- Low differential gain a. In theory a is expected to increase if the dimensionality of the electronic system of the active medium is reduced.

	Parameter	Value
Known Device Parameters	Confinement factor per layer	$\Gamma = 1\,\%$
	Waveguide width	$w = 4\ \mu m$
	Waveguide height	$d = 0.5\ \mu m$
	Active region height per QD-layer	$d' = 40$ nm
	Waveguide cross-section	$C = w \cdot d\ = 2 \times 10^{-12}\ m^2$
Known System Parameters for 40 Gbit/s OOK	Optical bandwidth	$B_O\ = 2$ nm
	Electrical bandwidth	$B_e\ = 50$ GHz
	Electrical SNR	SNR = 36
Fixed Parameters Obtained from Parameterization	Generation rate	$r = I\,/\,eV = 2.8 \times 10^{-32}\ s^{-1} m^{-3}$
	Wavelength	$\lambda_s\ = 1300$ nm
	Inversion factor	$n_{sp}\ = 1.5$
	Empirical tolerance factor	$\Delta P_{tol}\ = 12$ dB
	Normalized carrier lifetime	$\tau_e\,/\,N_t\ = 5.97 \times 10^{-33}\ s\ m^3$
	Normalized differential gain	$aN_t\ = 1.77 \times 10^4\ m^{-1}$

Table 4.3: Device and system parameters used for the optimization process and a QD SOA active region schematic are shown, [J2].

The lower limit of the IPDR$_f$ $P_1^{in,f}$ is given by (4.10) and (4.11). From these equations it can be derived that for a given unsaturated FtF gain $G_{f0} \gg 1$ the lower limit of the IPDR$_f$ $P_1^{in,f}$ is proportional to the inversion factor n_{sp}. The population inversion can be optimized by adapting the current accordingly, i.e., choosing it as high as possible. This will minimize the lower IPDR$_f$ limit.

Subsequently, the IPDR description is used to exemplarily derive design guidelines for a QD SOA with a FtF gain of $15...25$ dB and an IPDR$_f$ of $25...40$ dB. In general, we perform the study for different device lengths as well as for differing QD-layer numbers. From the discussion above we expect that longer devices with a lower confinement factor are preferable. All simulations are performed at 1.3 μm with a constant carrier generation rate $r = I/eV$. This way the effective carrier lifetime and the differential gain are constant. The current is always adapted because a variation of the QD-layer number l and device length changes the active volume $V = l \cdot d' \cdot w \cdot L$ (d' is the active region height per QD-layer and w is the waveguide width). The confinement is assumed to be 1 % per layer [23].

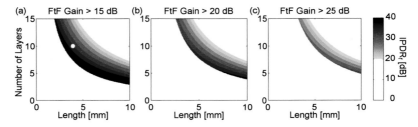

Fig. 4.12: Region for device parameters where the FtF gain and the IPDR$_f$ are larger than defined minimum values. For required values of the FtF gain, a low number of QD-layers and long waveguides provide the best IPDR$_f$. The circle indicates the IPDR$_f$ result measured with SOA 2 at 300 mA, [J2].

Firstly, the unsaturated FtF gain G_{f0} is calculated with (4.13) using the experimentally determined parameters τ_c/N_t, aN_t and r which are presented in Table 4.3. Once the unsaturated FtF gain G_{f0} is known, the saturation input power $P_{sat}^{in,f}$ can be calculated and subsequently using ΔP_{tol} the upper limit of the IPDR$_f$ $P_2^{in,f}$. Since the device is operated with a constant generation rate r also the inversion factor n_{sp} may be assumed to be constant which enables a calculation of the ASE output power $P_{ASE,f}$ with (4.11). Then, the lower IPDR$_f$ limit $P_1^{in,f}$ is calculated using (4.10). All parameters of the IPDR$_f$ description are summarized in Table 4.3. The IPDR$_f$ has been calculated for various gain values as a function of device length and number of QD layers. Fig. 4.12 shows the areas for which a particular minimal FtF gain (a) > 15 dB, (b) > 20 dB, (c) > 25 dB and minimal IPDR$_f$ (color-coded from $0...40$ dB) have been achieved. It can be seen that it is not possible to optimize device parameters in order to maximize the FtF gain and IPDR$_f$ simultaneously. In general, long devices with few QD-layers combine high gain and maximum IPDR$_f$. The IPDR$_f$ result measured with SOA 2 at 300 mA is indicated by the circle in Fig. 4.12(a).

4.4 IPDR for Advanced Modulation Format Signals

An ideal SOA for advanced modulation formats not only needs to properly amplify signals with various amplitude and power levels but also should preserve the phase relations between the symbols. Prior work on SOA has focused on DPSK [149]-[153], and DQPSK [154]. These modulation formats basically have a constant modulus and therefore naturally may be anticipated to be more tolerant towards SOA nonlinearities.

However, so-called M-ary QAM formats comprise both amplitude-shift keying (ASK) and PSK aspects. Thus, an ideal SOA should amplify such QAM signals with high amplitude and phase fidelity. In SOA gain changes and phase changes are related by the alpha factor α_H. Thus, one might assume that a low alpha-factor SOA is advantageous. As QD SOA tend to have lower alpha factors one might expect that they should outperform bulk SOA as amplifiers for advanced modulation formats.

In this section, it is show that SOA for advanced modulation formats primarily need to be optimized for a large IPDR. Additionally, an SOA with a low alpha factor offers advantages when modulation formats are not too complex, i.e., the number M of constellation points is not too high. The findings are substantiated by both simulations and experiments performed on SOA with different alpha factors for various advanced modulation formats. In particular it is shown that the IPDR advantage of a QD SOA with a low alpha factor reduces when changing the modulation format from BPSK (2QAM) to QPSK (4QAM) and it vanishes completely for 16QAM. This significant change is due to the smaller probability of large power transitions if the number M of constellation points increases. The smaller probability of large power transitions in turn leads to reduced phase errors caused by amplitude-phase coupling via the alpha factor.

4.4.1 Simulation of IPDR for Advanced Modulation Formats

In this subsection, the impact of the SOA's alpha factor on the amplification of advanced optical modulation format signals is investigated with simulations. Two SOA with identical performance in terms of unsaturated gain, saturation input power, noise figure, and SOA dynamics are used. The SOA only differ in their alpha factor. Simulations with alpha factors of 2 and 4 are performed for differentially phase-encoded and non-differentially phase-encoded data signals, respectively.

4.4.1.1 Models for Transmitter, SOA and Receiver

The simulation environment as shown in Fig. 4.13 consists of a 28 GBd transmitter (Tx), two virtual switches for either investigating the back-to-back (BtB) signal quality, or for simulating the influence of the SOA on the signal quality. The SOA model takes into account phase changes and ASE noise. The 28 GBd receiver (Rx) is either a direct receiver, a homodyne coherent or a differential (self-coherent) receiver, respectively. In the following each element of the simulation setup is discussed in more detail.

The transmitter (Tx) in Fig. 4.13 consists of a cw laser and an optical IQ-modulator. To achieve a realistic signal quality, the electrical SNR of the modulator input signal in the electrical domain is adjusted to 20 dB. A pulse carver is added in front of the IQ-modulator that either shapes the cw laser light to 33 % or 50 % RZ pulses, or just lets the cw light pass through for NRZ operation. The electrical data signals supplied to the optical modulator are low-pass filtered with a 3 dB bandwidth of 25 GHz. Jitter of 500 fs and rise and fall times of 8 ps are modelled to mimic realistic optical BtB data signals. This transmitter is used to generate signals with higher-order optical modulation formats such as (D)QPSK and 16QAM. For OOK and (D)BPSK data signals only the I-channel of the IQ-modulator is used. The output signal of the modulator is amplified and subsequently filtered by an optical band-pass filter. The in-fiber input power $P^{in,f}$ to the SOA can be adjusted with an attenuator.

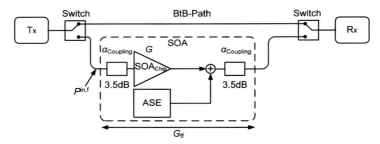

Fig. 4.13: Simulation environment for investigating the impact of the SOA alpha factor on signals with advanced optical modulation formats. The transmitter (Tx) generates OOK, (D)BPSK, (D)QPSK, or 16QAM data signals. Virtual switches define a reference path for back-to-back (BtB) simulations. The rate-equation based SOA model provides a chip gain G and takes into account fiber-to-chip losses $\alpha_{Coupling}$ of 3.5 dB per facet, gain-independent amplified spontaneous emission (ASE) noise, and phase changes. Depending of the transmitted data format, the receiver (Rx) is chosen, [J1].

The SOA is modelled following the numerical QD SOA model approach presented in *Section 2.3.3*. The model describes a QD SOA which is longitudinally subdivided into 20 segments. In each segment the evolution of the photon and carrier density along the propagation direction is governed by rate-equations for the photon number and the optical phase. Here, the ASE noise is simulated as white Gaussian noise added to the output of the SOA.

The SOA model parameters are shown in Table 4.4. The parameters are chosen (accordingly to the parameters of SOA 4 (*Section 3.2*)) to provide a FtF gain G_{ff} of 13.5 dB, a 3 dB in-fiber saturation input power $P_{sat}^{in,f}$ of −2 dBm and a FtF noise figure NF_{ff} of 8.5 dB. The estimated per-facet coupling loss $\alpha_{Coupling}$ is 3.5 dB.

To investigate the influence of the alpha factor on the amplification of signals with advanced modulation formats, we simulate two SOA with alpha factors 2 and 4, respectively. The FtF gain G_{ff} and the FtF noise figure NF_{ff} versus the in-fiber input power $P^{in,f}$ of the two SOA are shown in Fig. 4.14(a) and (b). The input signal is set to a wavelength of $\lambda_1 = 1554$ nm. The modulus $|\Delta\phi|$ of the phase changes of the SOA are plotted in Fig. 4.14(c). Large input powers cause carrier depletion. Thus, the gain is suppressed, and the phase change due the amplifier saturation increases with increasing input power. The device with the larger alpha factor shows stronger phase changes under gain suppression.

Parameter	Value
Current density	$J = 7.5 \text{ kA/cm}^2$
Area density of QD	$N = 8.5 \times 10^{14} \text{ m}^{-2}$
Number of QD layers (separated by spacers and wetting layers (WL))	$l = 6$
Amplifier length	$L = 1 \text{ mm}$
Thickness of active region	$t_{\text{wg}} = 0.125 \text{ μm}$
Width of active region	$w = 1.75 \text{ μm}$
Carrier lifetime (WL refilling time)	$\tau_{\text{c}} = 100 \text{ ps}$
Characteristic relaxation time(QD refilling time)	$\tau_1 = 1.25 \text{ ps}$
Relative line broadening (inhomogeneous/homogeneous)	$\gamma_{\text{inhom}} / \gamma_{\text{hom}} = 0.33$
Resonant cross-section (measure of photon-QD carrier interaction, describing the probability of stimulated radiative transitions)	$\sigma_{\text{res}} = 1.3 \times 10^{-19} \text{ m}^2$
Internal waveguide losses	$\alpha_{\text{int}} = 400 \text{ m}^{-1}$
Alpha-factor (Henry factor)	$\alpha_{\text{H}} = 2$ and $\alpha_{\text{H}} = 4$
Area density of WL states (WL states serve as a carrier reservoir for QD states)	$N_{\text{wl}} = 1.08 \times 10^{16} \text{ m}^{-2}$
Average binding energy (energy of QD electrons and holes relative to the WL bandedge)	$W_{\text{bind}} = 150 \text{ meV}$

Table 4.4: Parameter values of the QD SOA model (see *Section 2.3.3*) used in the simulations.

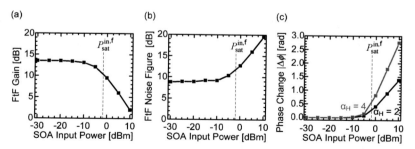

Fig. 4.14: Comparison of SOA characteristics for devices which only differ in the alpha factor. An SOA with an alpha factor of 2 (black) and an SOA with an alpha factor of 4 (blue) are used for the simulation. (a) FtF gain G_{ff} as a function of the SOA input power is shown. The unsaturated FtF gain G_{f0} is 13.5 dB at a wavelength of 1554 nm, and the 3 dB in-fiber saturation input power is −2 dBm. (b) FtF noise figure as a function of SOA input power. (c) Phase change $\Delta\phi \leq 0$ as a function of SOA input power. The SOA with larger alpha factor causes larger magnitudes $|\Delta\phi|$ if the SOA becomes saturated, [J1].

The receiver model depends on the modulation format. Basically, it comprises a noisy optical amplifier and a noise-free photoreceiver. Three receiver types are available: For direct detection, for coherent detection and for differential (self-coherent) detection. OOK-formatted (intensity encoded) signals are directly detected with a photodiode. The differentially phase-encoded DPSK and DQPSK formats are received with delay interferometer (DI) based demodulators followed by balanced detectors. Signals with non-differentially phase-encoded formats such as BPSK, QPSK and 16QAM are received using a homodyne coherent receiver comprising a noise-free local oscillator (LO), and balanced detectors for the in-phase and the quadrature-phase components, respectively.

4.4.1.2 Signal Quality Evaluation by Error Vector Magnitude, and Q^2

To estimate the signal quality of simulated (and measured) data signals after amplification with the SOA, we employ the error vector magnitude (EVM) for non-differentially phase-encoded data signals, and the Q^2 factor method for differentially phase-encoded data signals. With these data, we estimate the IPDR of the SOA.

Advanced modulation formats such as M-ary QAM encode the data in amplitude and phase of the optical electric field. The resulting complex amplitude of this field is described by points in a complex IQ constellation plane defined by the real part (in-phase, I) and imaginary part of the electric field (quadrature-phase, Q). Fig. 4.15(a) depicts a transmitted reference constellation point $E_{t,i}$ (●) and the actually received and measured signal vector $E_{r,i}$ (×), which deviates by an error vector $E_{err,i}$ from the reference. We use non data-aided reception and define the EVM for non-differentially phase-encoded data signals as the ratio of the root-mean-square (rms) of the EVM for a number of I_n received random symbols, and the largest magnitude of the field strength $E_{t,m}$ belonging to the outermost constellation point,

$$\mathrm{EVM}_m = \frac{\sigma_{err}}{|E_{t,m}|}, \; \sigma_{err}^2 = \frac{1}{I_n} \sum_{i=1}^{I_n} |E_{err,i}|^2, \; E_{err,i} = E_{r,i} - E_{t,i} \, . \tag{4.15}$$

The errors in magnitude and phase for the received constellation points are also evaluated separately. The EVM measures the quality of an advanced modulation format signal much the same way as it is customary with the Q^2 factor [155], [156].

With the measured EVM as in Fig. 4.15(b), the IPDR is defined as the range of input powers P^{in} into an SOA at which error-free amplification of a data signal can be ensured. The input power limits for error-free amplification are set by the EVM limit EVM_{lim} corresponding to a BER of 10^{-9} or 10^{-3}. The ratio of the corresponding powers P_2^{in} and P_1^{in} in Fig. 4.15(b) define the IPDR measured in dB, according to (4.9). The used EVM limits are 23.4 % for BPSK, 16.4 % for QPSK (indicating a BER = 10^{-9}), and 10.6 % for 16QAM (indicating a BER = 10^{-3}) [156].

Differential modulation formats such as DPSK or DQPSK encode information as phase difference between two neighboring bits. On reception, these phase differences are converted into an intensity change by using a delay interferometer demodulator. The signal quality of the obtained eye diagram is estimated by the Q^2 factor irrespective of the fact that demodulated phase noise is not necessarily Gaussian, and that therefore the inferred BER is inaccurate. The I and Q data of the DQPSK signals are evaluated separately and lead to virtually identical Q^2 factors.

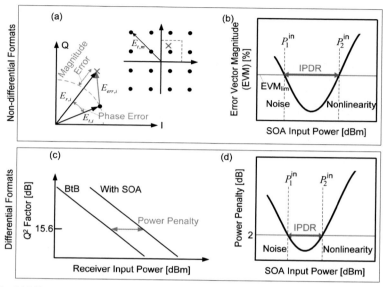

Fig. 4.15: Error-vector magnitude (EVM), power penalty (PP) and input power dynamic range (IPDR) for non-differential (QAM) and differential (DPSK, DQPSK) modulation formats. For QAM, subfigures (a) and (b) illustrate the EVM definition and the determination of the IPDR for given EVM_{lim}. For DPSK and DQPSK, subfigures (c) and (d) clarify what is meant with the power penalty for a given Q^2 of 15.6 dB, and how the IPDR is determined from a PP of 2 dB, [J1].

Fig. 4.15(c) the Q^2 factor as a function of receiver input power is presented schematically for the back-to-back case (BtB, without SOA) and for the case with SOA. The power penalty (PP) is the factor by which the power at the receiver input must be increased to compensate for signal degradations compared to the BtB case. In Fig. 4.15(d) the IPDR for DPSK and DQPSK is again defined according to (4.9), but this time by the logarithm of the ratio of SOA input powers P_2^{in} / P_1^{in} for which the power penalty is less than 2 dB at a signal quality of 15.6 dB (corresponding to a BER in the order of 10^{-9}).

4.4.1.3 Advantageous Modulation Formats for low Alpha–Factor SOA

In this subsection, it is shown by simulation that the use of a low alpha-factor SOA can have an advantage. The alpha factor mostly matters for simple phase-encoded signals. Fig. 4.16 shows for both SOA with alpha factors of 2 and 4 the respective EVM and the power penalties as a function of the SOA input powers for (a) BPSK, (b) QPSK, (c) 16QAM, (d) OOK, (e) DPSK and (f) DQPSK. In the case of the DQPSK modulation two variants are considered: The standard NRZ modulation technique which directly switches between different constellation points so that power transients occur, and a modulation format which maintains a constant signal envelope (red curve in Fig. 4.16(f) for $\alpha_H = 2$ and $\alpha_H = 4$), so that power transients are absent. The limiting EVM and the limiting power penalty are indicated by gray horizontal lines. The intersections of the EVM or PP curves with these horizontals define the limiting points for the IPDR$_f$.

The simulated IPDR$_f$ for modulation formats having power transitions between the constellation points reduces from BPSK (2QAM) to QPSK (4QAM) to 16QAM, and from DPSK to DQPSK. The IPDR difference for the SOA with different alpha factors is largest for the BPSK and the DPSK modulation format, which have the highest probability of large power transition. The power penalty as a function of SOA input power for OOK signal shows, as expected, no difference for the two SOA samples, Fig. 4.16(d). The results for constant-envelope DQPSK modulation exhibit a very low power penalty at high input powers, Fig. 4.16(f). This clearly demonstrates the strong influence of power transitions.

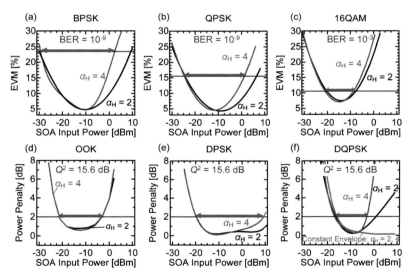

Fig. 4.16: Simulations illustrate an IPDR advantage for SOA with $\alpha_H = 2$ over SOA with $\alpha_H = 4$, if signals with advanced modulation format and large power transitions between the constellation points are amplified. (a)-(c) EVM as a function of SOA input power for BPSK, QPSK and 16QAM signals. The BtB values of the EVM are 4.8 % for BPSK, 5 % for QPSK and 7.5 % for 16QAM. (d)-(f) Power penalty as a function of SOA input power for OOK, DPSK and DQPSK signals. The red curve in subfigure (f) assumes a constant-envelope modulation and holds for both, $\alpha_H = 2$ and $\alpha_H = 4$. The IPDR$_f$ is indicated by red arrows, and the corresponding EVM$_{lim}$ and PP of 2 dB are shown by the gray horizontal lines, [J1].

Table 4.5 summarizes the IPDR$_f$ simulation results for both SOA types. The IPDR values are obtained assuming a certain BER limit (gray horizontal lines in Fig. 4.16), and for a specific evaluation method, i.e., power penalty (PP) or EVM. Further, the difference $IPDR_{f2} - IPDR_{f4}$ of the $IPDR_{f\alpha}$ for the devices with $\alpha_H = 2$ and with $\alpha_H = 4$ is specified.

Format	$-\log_{10}$ BER (PP, EVM)	$IPDR_{f\alpha}$ [dB]		IPDR Difference $IPDR_{f2} - IPDR_{f4}$ [dB]
		$\alpha_H = 2$	$\alpha_H = 4$	
33% RZ OOK	9 (PP)	19	19	0
NRZ DPSK	9 (PP)	~ 40	29	~11
NRZ DQPSK	9 (PP)	18	14	4
50% RZ DQPSK	9 (PP)	32	27	5
Const. Envelope DQPSK	9 (PP)	> 30	> 30	$IPDR_{2,4}$ indistinguishable
NRZ BPSK (2QAM)	9 (EVM)	39	32	7
NRZ QPSK (4QAM)	9 (EVM)	31	26	5
NRZ 16QAM	3 (EVM)	14	13	1

Table 4.5: Devices with lower alpha factor show larger IPDR for modulation formats with high probability of large power transitions. IPDR for various modulation formats for a symbol rate of 28 GBd for two SOA devices only differing in the alpha factor are shown. The evaluation method, i.e., PP or EVM and the corresponding BER limit are defined. The results of the IPDR difference are also presented. The advantage of a low alpha-factor device manifests in a large IPDR difference.

These results show that:

- For amplifying phase-encoded signals, low alpha-factor SOA are preferable, see $IPDR_{f2} - IPDR_{f4}$ in Table 4.5, e.g., rows "NRZ BPSK" and "QPSK".
- The influence of the alpha factor on high-order M-ary QAM signals reduces significantly, compare $IPDR_{f2} - IPDR_{f4}$ values in Table 4.5 rows "NRZ BPSK", "QPSK" with "NRZ 16QAM".
- As a general tendency, the IPDR reduces for increasing complexity of the optical modulation format: For a given average transmitter power the distance between constellation points reduces with their number, and the required OSNR increases. This moves intersection point P_1^{in} in Fig. 4.15 to higher powers. On the other hand, if the average power is increased to improve the OSNR, there is a larger risk of amplifier saturation, so that high-power constellation points move closer together. This would shift the intersection point P_2^{in} in Fig. 4.15 to lower powers.
- Long "1"-sequences of NRZ signals lead to a stronger carrier depletion than "1"-sequences of RZ signals, if the carrier recovery time (100 ps) is in the order of the pulse repetition rate (36 ps). This is the reason why in our case 50 % RZ DQPSK modulation leads to an $IPDR_f$ which is about 14 dB larger than that for NRZ DQPSK, Table 4.5.

4.4.1.4 Upper IPDR Limit as a Function of Alpha Factor and Symbol Rate

The evolution of the upper IPDR limit as a function of the SOA alpha factor is presented for 28 GBd NRZ DPSK and 28 GBd NRZ DQPSK in Fig. 4.17(a). Obviously, if the alpha factor increases from 2 to 8, the upper power limit of the IPDR reduces significantly for both modulation formats.

Depending on the transmitter, a phase change induced at the power transition time has not necessarily died out at the time of signal decision, so that the data phase is perturbed and errors occur. The strength of the phase change induced by the SOA and detected at the decision point depends on the required time to change between constellation points, and thus on the symbol rate and the transmitter bandwidth. If the time to change between constellation points (approximately 6 ps...30 ps for 50 GBd...10 GBd) is not significantly faster than the SOA phase recovery time (about 100 ps), severe phase changes will degrade the signal quality. Thus, only at higher symbol rates, no significant phase change from the SOA is expected indicating the benefit of a limited SOA recovery time. This behavior can be seen in Fig. 4.17(b) (adapted Tx, black curve). The figure shows the upper limit of the IPDR as a function of the symbol rate from 10 GBd to 50 GBd for two different Tx implementations. The adapted Tx curve describes a transmitter that electrical bandwidth and electrical rise / fall times are always adapted to the symbol rate. The Tx always shows amplitude transition which take about 1/3 of the symbol duration.

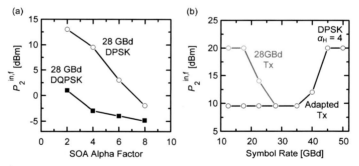

Fig. 4.17: Dependance of upper IPDR limit on SOA alpha factor and symbol rate. (a) Larger alpha-factor SOA significantly reduce the upper IPDR limit. Simulations are performed with 28 GBd DPSK and 28 GBd DQPSK signals. (b) Benefit of limited SOA phase recovery speed. In case of a Tx designed for a specific symbol rate (adapted Tx, black curve), the upper IPDR limit increases with increasing symbol rate. At very high symbol rates, the amplitude transition is too fast, so that the SOA phase change does not follow. If a high-speed Tx (always 28 GBd Tx) is used also to encode lower symbol rates, the SOA phase recovery is sufficiently fast to follow the pattern. Thus, the upper limit of the IPDR increases with a decrease in the symbol rate (red curve). The target BER is 10^{-9}, the upper IPDR limit is determined with a power penalty of 2 dB for DPSK data signals and an SOA with an alpha factor of 4 (upper limits of the IPDR exceeding 20 dBm are not further investigated, and thus the power limit is indicated in the figure at 20 dBm).

On the contrary, if the 28 GBd Tx is always used, also at lower symbol rates (electrical signal applied to the optical modulator has always 8 ps rise / fall times irrespective of the symbol rate), the SOA induced phase changes are approximately independent of the symbol rate. Thus, the upper limit of the IPDR increases if the symbol rate decreases. This is due to the fact that the SOA phase recovery upon a fast amplitude transition is now sufficiently fast at a low symbol rate. The simulations are performed for NRZ DPSK with an alpha factor of 4.

4.4.2 Measurements of QAM Formats

To verify the prediction from the simulations two SOA devices are tested with 20 GBd BPSK, QPSK and 16QAM signals. Both SOA are very similar in terms of gain, noise figure, saturation input power as well as dynamics. However, the devices differ in their alpha factors since actually different structures are used, i.e., a bulk and a QD SOA.

4.4.2.1 QD and Bulk SOA Characteristics

The experiments are performed with a 1.55 µm QD SOA (SOA 4) and a 1.55 µm low optical confinement bulk SOA (SOA 6) [49]. They have been operated with the same current density. Fig. 4.18(a) shows that FtF gain, FtF noise figure and in-fiber saturation input powers are indeed comparable. The gain peak of both devices is around 1530 nm, and the −3 dB bandwidth is 60 nm each.

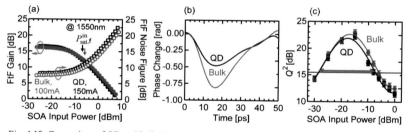

Fig. 4.18: Comparison of QD and bulk SOA characteristics. (a) FtF gain, FtF noise figure and in-fiber saturation input powers for a 1.55 µm QD SOA (black) and bulk SOA (blue). For equal current densities all characteristics are comparable. (b) Phase response in relation to an 8 ps wide impulse. The bulk SOA shows 1.7 times the peak-to-peak phase change of the QD SOA. (c) Q^2 factor for amplification of a 43 Gbit/s RZ OOK data signal for different device input powers. Since the dynamic range (IPDR$_f$ indicated by red arrow, gray horizontal line is $Q^2 = 15.6$ dB) of both SOA is almost identical, the device performance differs only in the alpha factor, [J1], [C19].

Fig. 4.18(b) shows the phase response of the QD and bulk SOA to an 8 ps wide impulse with an average input power of +7 dBm. The phase change has been measured with a frequency resolved electro-absorption gating (FREAG) technique based on linear spectrograms [123], see *Chapter 3*. The bulk SOA shows 1.7 times higher phase changes than the QD SOA. Therefore, the ratio of the alpha factors is $\alpha_{H,bulk} / \alpha_{H,QD} = 1.7$. In Fig. 4.18(c) the signal quality (Q^2) of a 43 Gbit/s RZ OOK data signal with varying SOA input power is shown after amplification with the QD and bulk SOA. The IPDR$_f$ for the target Q^2 factor of 15.6 dB is around 22 dB for both amplifiers. From these findings we conclude that the devices are comparable with respect to their gain recovery times, and that the overall performance only differs in their phase responses. This fact enables a comparison for advanced modulation format signals in terms of the alpha factor only.

4.4.2.2 Multi-format Transmitter and Coherent Receiver

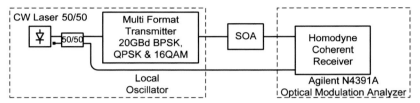

Fig. 4.19: Experimental setup, comprising a software-defined multi-format transmitter encoding 20 GBd BPSK, QPSK and 16QAM signals onto an optical carrier. The signal power level is adjusted before launching it to the QD or bulk SOA (DUT). The optical modulation analyzer receives, post-processes, and analyzes the data, [J1].

The IPDR for amplification of NRZ BPSK, NRZ QPSK and NRZ 16QAM data signals has been studied by evaluating the EVM. The experimental setup (Fig. 4.19) comprises a software-defined multi-format transmitter [157] encoding the data onto the optical carrier at 1550 nm, the devices under test (DUT) and a coherent receiver (Agilent N4391A Optical Modulation Analyzer (OMA), generation and reception in *Appendix A.1, A.3*). The symbol rate is 20 GBd resulting in 20 Gbit/s BPSK, 40 Gbit/s QPSK and 80 Gbit/s 16QAM signals. The power of the signal is adjusted before launching it into the SOA. After amplification, we analyze EVM as well as magnitude and phase errors. The OMA receives, post-processes, and analyzes the constellations.

4.4.2.3 Large IPDR with Low Alpha-factor SOA for Low-order QAM Formats

In Fig. 4.20(a)-(c) the EVM for the different modulation formats is depicted as a function of the SOA input power. Fig. 4.20(a) shows for BPSK modulation an IPDR$_f$ exceeding 36 dB with around 8 dB enhancement for the QD SOA compared to the bulk SOA. Fig. 4.20(b) shows for QPSK modulation an IPDR$_f$ of 29 dB with an improvement of 4 dB for the QD SOA. The IPDR$_f$ for 16QAM is 13 dB, and shows no difference between both amplifier types, see Fig. 4.20(c).

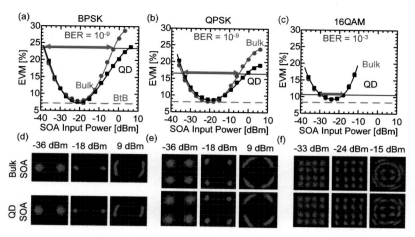

Fig. 4.20: EVM for different modulation formats and two types of SOA versus input power. (a) Low alpha-factor QD SOA shows an IPDR$_f$ enhancement of 8 dB compared to bulk SOA for BPSK modulation. In both cases the IPDR$_f$ exceeds 36 dB. (b) IPDR$_f$ enhancement at QPSK is reduced, but still 4 dB. An IPDR$_f$ of 29 dB is found. (c) No difference is found at 16QAM. The IPDR$_f$ for both devices is 13 dB. The BtB EVMs are indicated by the red dashed lines. The IPDR are shown by the red arrows, and the gray horizontal lines represent the EVM$_{lim}$. (d)-(f) Constellation diagrams for various SOA input powers which are associated with the respective subfigure (a)-(c) immediately above. Bulk SOA (upper row) and QD SOA (lower row) are compared, [J1].

In addition, in Fig. 4.20(d)-(f) constellation diagrams for bulk SOA and QD SOA are presented below the respective EVM subfigures for the three modulation formats, and for three different input power levels. For low input powers the constellations points are broadened by ASE noise. For optimum input powers the constellation points have almost BtB quality. For large input powers the signal quality again reduces. Obviously, the limitation for BPSK and QPSK stems from phase errors, whereas the limitation for the 16QAM signal stems from both, amplitude and phase errors. The phase errors with the PSK formats are larger for the bulk SOA than for the QD SOA.

It has already been shown that EVM is an appropriate metric to estimate the BER and describe the signal quality of an optical channel limited by additive white Gaussian noise [156]. Here, the EVM is also used to estimate the BER if the signal quality is limited by nonlinear distortions. Amplitude and phase errors induced by the SOA causes an EVM which underestimates the BER. The discrepancy in terms of SOA input power is in the range of 2..3 dB. Thus, the EVM is used to obtain a reliable tendency of the IPDR, see *Appendix A.6*.

To study the IPDR limitations for low and high input power levels, the magnitude and phase errors (Fig. 4.21) relative to the BtB magnitude and phase values are evaluated. For low input power, Fig. 4.21(a)-(c) shows that magnitude and phase errors decrease with increasing input power. No difference can be seen between bulk and QD SOA. The behavior of the SOA samples differs, however, for large input powers. For BPSK and QPSK encoded signals the amplitude is virtually error-free, whereas the phase error increases with increasing input power. For the 16QAM signal, both, magnitude and phase errors contribute to the EVM.

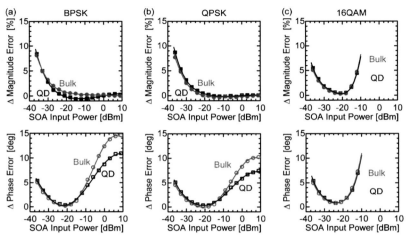

Fig. 4.21: Magnitude error and phase error increase as compared to BtB measurements. The degradation for low input power is due to OSNR limitations. The upper limit is due to phase errors for (a) BPSK and (b) QPSK. Magnitude errors are insignificant. (c) At 16QAM the phase error is accompanied by gain saturation inducing magnitude errors, [J1].

4.4.3 Measurements of Differential PSK Modulation Formats

In addition to QAM formats, it is interesting to investigate in experiments also non-differential PSK formats, because of the different detection method, see *Appendix A.3*. Further in this subsection also the multi-wavelength performance of SOA for PSK signals is tested.

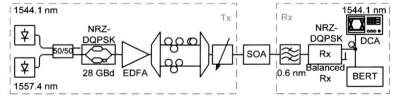

Fig. 4.22: Experimental setup. Two 28 GBd NRZ DQPSK channels are equalized and de-correlated using 0.5 m of single-mode fiber. The average power of both channels is varied and launched into a bulk or QD SOA. A single channel is selected, amplified and demodulated in a DQPSK receiver (Rx), consisting of a delay interferometer (DI) followed by a balanced detector. The electrical signal is then analyzed using a digital communications analyzer (DCA) and a bit-error ratio tester (BERT), [J6].

The power penalty caused by an SOA for NRZ DQPSK data is investigated using the experimental setup shown in Fig. 4.22. It comprises two data signals at 1554.1 nm (ch. 1) and 1557.4 nm (ch. 2), which are de-correlated by 69 bit (generation in *Appendix A.1*). The power levels of both channels are adjusted to be equal before launching them into the SOA. After amplifying both data signals in the SOA, the 1557.4 nm channel is blocked by an optical band-pass filter while the BER of the remaining data channel is analyzed. The DQPSK Rx consists of a DI (see *Appendix A.2*) based demodulator followed by a balanced detector and a bit-error ratio tester (BERT, reception in *Appendix A.3*). In the experiment, no data encoder circuit was employed. To allow bit-error ratio measurements, the error detector is programmed with the expected data sequence, which allows a pseudo-random bit sequence (PRBS) length of up to 2^9-1.

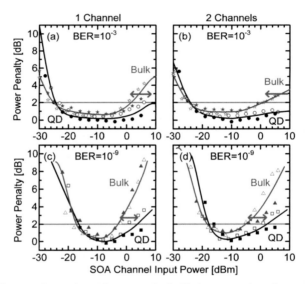

Fig. 4.23: Power penalty vs. channel input power levels. The input power dynamic range (IPDR$_f$) is defined as the range of input power levels with less than 2 dB power penalty compared to the back-to-back case. Red arrows indicate the IPDR enhancement of the QD SOA (SOA 4, black) over the bulk SOA (SOA 6, blue). (a), (b) QD SOA improve the IPDR at a BER of 10^{-3} by 5 dB and > 10 dB compared to bulk SOA for one and two 28 GBd NRZ DQPSK channels, respectively. (c), (d) For a BER of 10^{-9}, the QD SOA IPDR is enhanced by 5 dB. The filled symbols correspond to the I-channel, whereas the open symbols represent the Q-channel. The total bit rate is 56 Gbit/s for one channel and 112 Gbit/s for two channels, respectively, [J6].

The IPDR for amplification of one and two 28 GBd NRZ DQPSK channels is studied by evaluating the power penalty for two selected bit-error ratios of BER = 10^{-9} and BER = 10^{-3}. Fig. 4.23 shows the power penalty as a function of the SOA channel input power for one and two channels for a specific BER. Fig. 4.23(a) and (b) show an IPDR$_f$ improvement of 5 dB for the single channel and > 10 dB for the two channel case for a BER of 10^{-3}, respectively. Fig. 4.23(c) and (d) show around 5 dB IPDR improvement for the QD SOA (SOA 4) compared to the bulk SOA (SOA 6) for one and two NRZ DQPSK channels at a BER of 10^{-9}. The filled symbols correspond to the I-channel, whereas the open symbols represent the Q-channel. The QD SOA exhibits a large IPDR of around 20 dB for BER = 10^{-9} and exceeds 30 dB for BER = 10^{-3}. We attribute the increased $IPDR_f$ in the two-channel case to the effect of XGM, which reduces the gain for each channel. This leads to larger saturation power levels, less patterning effects and thus smaller power penalties. The results are summarized in Table 4.6.

		1 Channel		2 Channels	
	BER	10^{-3}	10^{-9}	10^{-3}	10^{-9}
	QD	> 32	19.5	> 36	19.4
$IPDR_f$ [dB]	Bulk	26.9	15.0	25.7	13.4
	Enhancement	> 5.1	4.5	> 10	6.0

Table 4.6: Input power dynamic range (IPDR) at 2 dB power penalty for bulk and QD SOA. In all cases, the QD SOA (SOA 4) shows a significant IPDR enhancement compared to a conventional bulk SOA (SOA 6), [J6].

To illustrate this advantage of QD SOA over bulk SOA, Fig. 4.24 shows the observed eye diagrams of the demodulated 28 GBd NRZ DQPSK Q-channel for high SOA input power levels of 6 dBm. The comparison of the back-to-back eye diagram in Fig. 4.24(a) for high input power to the QD SOA, eye diagrams in Fig. 4.24(b), shows no signal degradation at optimum receiver sensitivity. In contrast, for high input power to the bulk SOA, the eye opening in Fig. 4.24(c) is reduced, and the signal quality is significantly degraded.

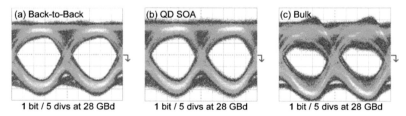

Fig. 4.24: Demodulated NRZ DQPSK eye diagrams for high SOA input power levels of 6 dBm. (a) Back-to-back eye diagram of 28 GBd Q-channel at optimum receiver sensitivity. (b) No signal degradation for QD SOA high input powers. (c) Reduced eye opening and signal quality degradation for bulk SOA, [J6].

As is well known from OOK formats, SOA can introduce strong amplitude distortions like overshoots [158] and patterning effects [159] when operated in saturation. For ideal phase-encoded signals, these effects should be strongly suppressed [149].

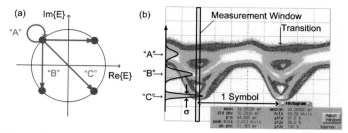

Fig. 4.25: Amplitude distortions of the NRZ DQPSK envelope in QD and bulk SOA. (a) Possible transitions in the constellation diagram. For a better long-term stability of the transmitter, only some transitions ("B", "C") are generated using amplitude modulation, instead of employing pure phase modulation (all states on the circle). (b) Example of the power envelope eye diagram for a 28 GBd NRZ DQPSK signal, showing all possible transitions. A histogram is taken at the transition in a 1.6 ps time window. Assuming a Gaussian distribution, the mean value and standard deviation are investigated as a function of the device input power for both device types, [J6].

Fig. 4.25(a) shows all possible transitions in the constellation diagram of DQPSK signals. Instead of pure phase modulation (all states on the circle), most practical implementations generate some transitions ("B", "C") using amplitude modulation. Fig. 4.25(b) shows an example of the power envelope eye diagram for a 28 GBd NRZ DQPSK signal with constant power at the decision point in the middle of the bit slot. All expected transitions are observed. A histogram is taken at the transition in a 1.6 ps time window. To investigate the influence of the amplitude effects for both SOA types, the power envelope of the 28 GBd NRZ DQPSK signal is analyzed as a function of the channel input power. Assuming a Gaussian distribution, mean value and standard deviation are calculated.

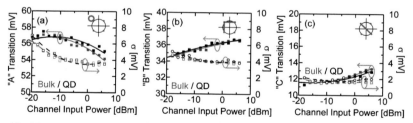

Fig. 4.26: Amplitude transitions in bulk (blue) and QD SOA (black): Dependance of mean values (■, filled symbols) and standard deviations (□, open symbols) as a function of the channel input power. For all three possible transitions, bulk and QD SOA show an identical behavior. Measured at optimum receiver input power, the standard deviations are unaffected by input power levels above −10 dBm. No difference between bulk and QD SOA is observed with respect to the signal amplitude, [J6].

In DQPSK systems with direct detection, the signals are received using a delay interferometer (DI) followed by a balanced receiver. Fluctuations of the received power as well as deviations from ideal phase transitions can strongly degrade the signal.

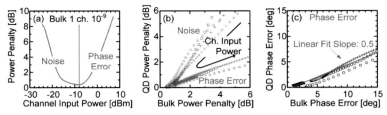

Fig. 4.27: Comparison of QD and bulk power penalty for all channels and BER. (a) Bulk SOA 1 ch. power penalty for BER = 10^{-9} as an example. The power penalty curves are marked according to the corresponding limits of the IPDR. For low input power levels, the DQPSK signal is degraded by noise (green). For high input powers, saturation of the SOA induces phase errors (red). (b) Power penalty for QD SOA vs. power penalty for bulk SOA when increasing the channel input power for BER = 10^{-9} ($\circ$: 1 ch., +: 2 ch) and BER = 10^{-3} ($\square$: 1 ch., $\times$: 2 ch.). The penalty is attributed to either noise or phase errors. All measurements shown in Fig. 4.23 displayed a similar ratio of the penalties. The largest difference between the samples arises for high input powers. (c) The power penalty for high input powers can be related to an effective phase error of the delay interferometer by the relation presented in [160]. The slope of a linear fit is 0.5, which actually corresponds to the ratio of the respective alpha factors. The very good agreement between the calculated and measured ratio of the alpha factors provides the explanation of the advantage of QD SOA in terms of IPDR: The smaller alpha factor of QD SOA reduces the phase impairments.

Fig. 4.26 shows the dependance of mean values (■, filled symbols) and standard deviations (□, open symbols) of bulk (blue) and QD SOA (black) as a function of the channel input power. In all cases, bulk and QD SOA behave identically. Measured at optimum receiver input power, the standard deviations are unaffected by input power levels above −10 dBm. No difference between bulk and QD SOA is observed with respect to the signal amplitude which is similar to the results from *Section 4.4.2* for 4QAM (QPSK).

Direct detection receivers for differential phase-encoded signals are particularly susceptible to errors caused by deviations from the ideal phase transitions. For a power penalty less than 2 dB the phase error at the DI must be less than 10° [160], [161]. Due to the fact that typical NRZ DQPSK transmitters show a fast amplitude transition if a phase change occurs, SOA can induce errors by amplitude and phase fluctuations [143].

As an example, the bulk SOA (SOA 6) 1 ch. power penalty for BER = 10^{-9} is depicted in Fig. 4.27(a). The power penalty curves are marked according to the corresponding limits of the IPDR. For input power levels below −10 dBm, the DQPSK signal is limited by noise. For input power levels above −10 dBm, saturation of the SOA induces phase errors. The main difference between the samples arises for high input powers. Therefore, the phase limitations on the DQPSK signal performance is studied.

Fig. 4.27(b) shows the penalty for the QD SOA vs. the power penalty for the bulk SOA when increasing the channel input power levels for BER = 10^{-9} ($\circ$: 1 ch., +: 2 ch) and BER = 10^{-3} ($\square$: 1 ch., $\times$: 2 ch.). The penalty is attributed to either noise or phase errors. All measured data which have been displayed in Fig. 4.23 resulted in a similar penalty ratio.

The influence of small phase errors is equivalent to the effect of deviations from the optimum operating point of the delay interferometer. For demonstrating this influence on the power penalty, we calculate the equivalent phase misalignment of a DI that would lead to the actually measured power penalty [160]. The absolute value of this phase error contains large uncertainties due to the fact that the phase error probability density is unknown. However, bulk and QD SOA are operated under identical conditions, and the calculated phase errors of the bulk SOA can be therefore compared to the phase errors of the QD SOA. As the biggest phase error will determine the bit-error ratio, the calculated values give an estimate of the worst-case phase error.

Fig. 4.27(c) compares the calculated equivalent phase errors for the bulk SOA to the equivalent phase errors for the QD SOA. A linear fit of the data shows a slope of 0.5. Assuming that the observed power penalty can be completely attributed to phase errors, this slope gives the ratio of the alpha factors of both devices. It is in good agreement with the results extracted from the independently measured P2P phase changes using the FREAG technique, see Fig. 4.18.

In Table 4.7 the experimentally obtained results of the IPDR_f difference for both SOA ($\text{IPDR}_{f,\text{QDSOA}} - \text{IPDR}_{f,\text{bulkSOA}}$) are summarized and compared to the simulation result (*Section 4.4.1*) for various modulation formats (also including the measurement results from *Section 4.4.2.3*). The IPDR_f differences obtained from simulations predict the results of the measurement. The evaluation method, i.e., EVM or power penalty and the corresponding BER limits are denoted.

Format	$-\log_{10}$ BER	IPDR Difference	
		$\text{IPDR}_{f,\text{QDSOA}} - \text{IPDR}_{f,\text{bulkSOA}}$ [dB]	$\text{IPDR}_{f2} - \text{IPDR}_{f4}$ [dB]
	(PP, EVM)	Measurement	Simulation
NRZ DQPSK	9 (PP)	5	4
NRZ BPSK	9 (EVM)	8	7
NRZ QPSK	9 (EVM)	4	5
NRZ 16QAM	3 (EVM)	0	1

Table 4.7: Comparison of measurement and simulation results for the difference of the IPDR for the lower alpha-factor SOA and the higher alpha-factor SOA for various modulation formats. Measurements and the simulations show the same tendency in spite of the fact that the symbol rate had to be reduced for the measurement from 28 GBd (as assumed for the simulations) to 20 GBd for the QAM data signals due to limitations in the available equipment.

4.5 IPDR of Small–Signal Modulated Data Signals

Radio-over-fiber systems are used for the transport and distribution of radio-frequency (RF) signals such as used in wireless local area networks (WLAN), WiMAX, or mobile communications over a fiber infrastructure [162]. Currently, the convergence of gigabit passive optical networks (GPON) and RoF systems is discussed, see *Section 1.4*. For reach extended GPON, SOA have emerged as cost effective and viable amplifiers. However, with respect to RoF signals, this leads to the question of the SOA's optical IPDR, as customers are located at different distances to the SOA, and whether a large electrical power dynamic range (EPDR) can be maintained. Actually, a large EPDR is desirable in the face of RF path fading (exceeding 30 dB), movement of customers inside radio cells, and RF-to-optical conversion inefficiencies. Up to now, a study of $IPDR_f$ and EPDR for SOA in such RoF networks is still missing. RoF signals are intensity modulated onto an optical carrier and launched into the SOA which amplifies these small-signal modulated signals.

In this section, it is shown that the SOA 2 (QD SOA) provides a huge 30 dB input optical and 30 dB electrical power dynamic range, respectively. EPDR can be optimized at the expense of $IPDR_f$ and vice versa. Experiments have been performed with QPSK and 16QAM RoF signals at 10 MBd.

4.5.1 IPDR and EPDR in Radio-over-Fiber Networks

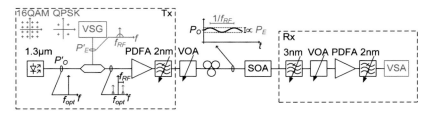

Fig. 4.28: Experimental setup with transmitter (Tx) and receiver (Rx) for determination of the optical input power dynamic range and the electrical power dynamic range of the QD SOA (SOA 2) at 1.3 μm. QPSK respectively 16QAM signals from a vector signal generator (VSG) at 2 GHz and 5 GHz with 10 MBd are modulated onto an optical carrier. P_O is the optical power launched into the SOA. The modulation depth of P_O induced by the RF power P_E after the Tx is illustrated in the inset above the SOA. The signals are detected with a pre-amplifier receiver and the error vector magnitude is evaluated with a vector signal analyzer (VSA), [C9].

To determine the optical $IPDR_f$ and the electrical EPDR, the QD SOA is placed between the Tx and the Rx of the experimental setup as shown in Fig. 4.28. The transmitter comprises a 1.3 μm cw laser (optical output power P'_O, centre frequency f_{opt}) which is fed into a Mach-Zehnder modulator (MZM) biased at its quadrature point. Digitally modulated QPSK and 16QAM signals at a symbol rate of 10 MBd are generated by a vector signal generator (VSG) at RF carrier frequencies of $f_{RF} = 2$ GHz and 5 GHz. The RF signals with power P'_E from the VSG are fed into the MZM. The RF signals reside as sidebands on the intensity-modulated optical carrier. The optical signal is then amplified by a praseodymium doped fiber amplifier (PDFA) and the optical power levels are adjusted by a variable optical attenuator (VOA) in front of the SOA 2 (QD SOA).

For the remainder of this subsection, P_O denotes the power of the optical carrier from the Tx into the SOA 2. Moreover, P_E denotes the modulation depth induced by the electrical power of the RF signal that is modulated onto this optical carrier. The optical power into the pre-amplifier receiver (Rx) is kept constant at −15 dBm in order to maintain the same sensitivity. The RoF signals are demodulated by a vector signal analyzer (VSA) and the EVM normalized to the average signal power is measured. An EVM of 16.7 % for QPSK and 7.5 % for 16QAM, respectively, is calculated to correspond to BER of 10^{-9} (error-free operation) [155]. The minimum required P_E for error-free operation without the SOA is −32 dBm at 2 GHz and −30.5 dBm at 5 GHz for QPSK and −24 dBm at 2 GHz and −23.5 dBm at 5 GHz for 16QAM, respectively.

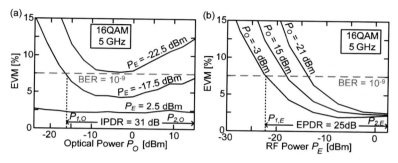

Fig. 4.29: Map of EVM for measured combinations of P_O and P_E operating points. (a) EVM as a function of optical power into the QD SOA. The larger the RF power level P_E, the larger the IPDR$_f$. (b) EVM as a function of RF power P_E. The largest EPDR exceeding 30 dB is offered if the optical input power P_O is near the SOA saturation input power, which here is around $P_O = -3$ dBm, [C9].

EVM measurements are carried out for both QPSK and 16QAM signals at 2 GHz and 5 GHz. The measurement range for P_E and for P_O are − 30 dBm < P'_E < +5 dBm (corresponding to a modulation depth of the optical carrier between 0.1 % and 30 %) and − 24 dBm < P_O < + 15 dBm, respectively. Fig. 4.29(a) shows a map of EVMs as a function of P_O for different P_E for the 16QAM signal on the 5 GHz carrier. The dashed line indicates the EVM limit for error-free operation at a BER of 10^{-9}. At low P_O, the signal performance degrades as the device is operated below the OSNR limit. At high P_O, the device is operated in gain saturation leading to a higher EVM. For subsequent use, the IPDR$_f$ is defined as the power range within which error-free amplification is guaranteed for a particular P_E. E.g., in Fig. 4.29(a) the lower ($P_{1,O}$) and upper ($P_{2,O}$) limits provide an IPDR$_f$ > 30 dB for P_E > −17.5 dBm (limits of the IPDR$_f$ are re-named in this section to distinguish optical and the electrical power levels).

Fig. 4.29(b) shows a map of EVM as a function of P_E for different P_O. The EVM decreases for increasing P_E. Best performance is achieved for P_O close to the SOA saturation input power. The EPDR is defined in analogy to the IPDR$_f$ as the operation range within error free operation is obtained for a given P_O. While the SOA provides gain, i.e., budget extension, it also induces signal degradation. This degradation may be overcome by modulating the RF signal at a higher power. This additional power is defined as the RF power penalty.

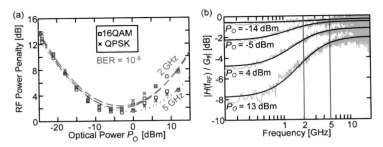

Fig. 4.30: (a) Additional RF power needed in order to maintain the 10^{-9} BER level in the system with an SOA. The RF power required is plotted as a function of the optical input power levels for QPSK and 16QAM at 2 GHz and 5 GHz. At low optical power, the RF power penalty increases as the device is operated below the optical signal-to-noise ratio limit. At high optical power, the device is operated under gain saturation which also increases the penalty. (b) Modulation response of QD SOA for different power levels showing the origin of the RF carrier frequency related penalty at higher optical input power levels P_O, [C9].

It is plotted in Fig. 4.30(a) for QPSK and 16QAM at both 2 GHz and 5 GHz. It is observed to be independent of the modulation format. The minimum RF power penalty of 1 dB (2 dB) for the 5 GHz (2 GHz) signals occurs at $P_O = -5$ dBm. At lower P_O, the signal performance is degraded by noise and the power penalty increases independently of the carrier frequency.

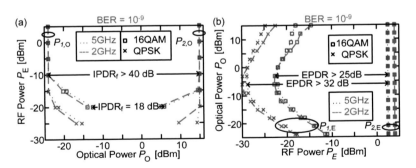

Fig. 4.31: Map of (a) $IPDR_f$ operation ranges for various P_E and (b) EPDR operation ranges for various P_O for the considered RoF signals. In (a) an ultra-large $IPDR_f > 40$ dB is observed for RF powers > -15 dBm for QPSK and > -10 dBm for 16QAM. (b) Lower and upper limit of the EPDR for all optical power levels. Maximum EPDR > 32 dB are observed for QPSK and EPDR > 25 dB for the more demanding 16QAM at -5 dBm optical power, [C9].

However, at high P_O the degradation is stronger for the 2 GHz signals than for 5 GHz signals. This is mostly due to the frequency dependance of the gain in saturation. This can be deduced from the modulation response $|H_m(f_{RF})| = |H(f_{RF}) / G_{fl}|$ of the SOA, see *Section 2.2.7*.

Fig. 4.30(b) shows the magnitude of the modulation response from 0.1 GHz to 20 GHz for different P_O at 1.3 μm. For low P_O, the modulation response is flat, whereas for high P_O the modulation response has high-pass characteristic. Under deep gain saturation and for low RF frequencies (0.1 GHz to 1 GHz), the SOA gain can follow the small signal modulation [163]. Further explanations are reported in *Section 2.2.7*.

Fig. 4.31(a) shows the IPDR$_f$ with its lower ($P_{1,O}$) and the upper ($P_{2,O}$) limits over P_E. An ultra-large IPDR$_f$ > 40 dB is observed for sufficiently high P_E. With an IPDR$_f$ > 40 dB, the QD SOA (SOA 2) works well independently of the distance to the customer, i.e., with a weak or a strong RoF signal. For weak P_E the ideal IPDR$_f$ range is small and the optical power P_O into the SOA must be chosen reasonably large. As a consequence the SOA should be used as an in-line amplifier in the upstream of an extended GPON with, i.e., 1:32 split ratios.

Fig. 4.31(b) shows the lower ($P_{1,E}$) and the upper ($P_{2,E}$) limits for P_E and thus the EPDR for error-free operation for all measured P_O. Maximum EPDR > 32 dB for QPSK and EPDR > 25 dB for the more demanding 16QAM are observed at the ideal P_O. Thus, in scenarios where a large EPDR is needed, the SOA should be operated around $P_O = -5$ dBm.

4.6 Linear SOA Applications in Future Optical Networks

In this section, network applications of linear SOA offering a large IPDR are presented. First, the loss budget increase induced by an SOA reach extender in a WDM/TDM GPON is investigated with a QD SOA. Four downstream channels with a data rate of 2.5 Gbit/s each are amplified in a 1.55 μm QD SOA. In the upstream direction two channels at a data rate of 622 Mbit/s each are amplified using a 1.31 μm QD SOA. A large input power dynamic range exceeding 40 dB is demonstrated which is needed to handle user length and user power variations. Each channel serves a reach of 60 km and is branched out with a 1:32 splitter. A total loss budget in the order of 45 dB is supported.

Second, the burst mode capability of SOA is addressed by measurements of the amplification of bursts with a length of 100 μs with 1.25 Gbit/s OOK data signals for different power levels.

Third, rules for the range of optical channel power of a cascade of SOA to guarantee the functionality of a metro and access network comprising several ROADM are provided. Experiments in a recirculating fiber loop with a length of 17 km with one QD SOA are performed with intensity-modulated (IM) direct-detected (DD) OFDM-BPSK, OFDM-QPSK as well as OOK data formats. It is found that the NRZ OOK format is better in terms of achievable distance and input power dynamics compared to the OFDM-BPSK/QPSK formats.

4.6.1 SOA Reach extender for PON for OOK Signals

Typical extended reach WDM/TDM GPON architecture (see Fig. 4.32, and Fig. 1.4) requires optical amplification which is provided in this approach by QD SOA. The 1.55 μm QD SOA (SOA 4) supports the needs of a downstream reach extender with a gain larger 10 dB and a reasonable low noise figure. The requirements for the amplifying SOA in the upstream path are a high IPDR and a large burst-mode tolerance due to the burst nature of upstream traffic. The 1.31 μm QD SOA (SOA 2) meets the needs by showing a noticeable IPDR.

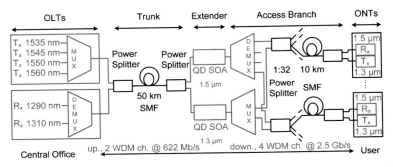

Fig. 4.32: Extended WDM/TDM GPON testbed with 4 downstream and 2 upstream channels, each serving 32 subscribers with 60 km reach using QD SOA technology for amplification, [C25].

The potential of QD amplifiers to compensate losses for achieving an extended reach is explored. Since the losses in the trunk (Att.1) and the losses in the access branch (Att.2) are interrelated, we determined first the acceptable losses after the QD SOA as a function of the given losses in front of the QD SOA, both for upstream and downstream (see Fig. 4.33). The losses are acceptable as long as the quality factor Q^2 is better than 15.6 dB.

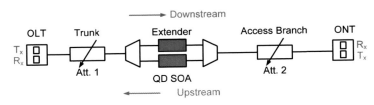

Fig. 4.33: Up-/ downstream path to identify the reach extension offered by QD SOA in a GPON. Two attenuators represent the trunk losses and the access branch losses given by the WDM filter, splitter and branch losses, respectively, [C25].

Fig. 4.34(a) depicts the access branch loss versus the acceptable loss in the trunk for a 1.31 μm QD SOA (blue curve) and Fig. 4.34(c) shows the dependance of loss in the trunk line as a function of the acceptable access branch loss for a 1.55 μm QD SOA (red curve). The results with QD SOA are compared to the case without amplification (black curves). For high losses in the order of 20 dB before the amplifier the QD SOA extend the loss budget up to 17 dB and 14 dB for the downstream and the upstream, respectively. At these optimum operating points (see circles ○ in Fig. 4.34), the total acceptable losses after the amplifier exceeds 30 dB for downstream and upstream. Therefore, state-of-the-art QD SOA support a total loss budget exceeding 50 dB. Dispersion has been neglected in this study due to the low bit rates. The data in Fig. 4.34 are measured with signals at 2.5 Gbit/s NRZ OOK (1550 nm) and at 622 Mbit/s NRZ OOK (1310 nm). A power of +5 dBm for each channel is launched at the optical line terminal (OLT) and optical network terminations (ONT), respectively. The receiver sensitivity is −33 dBm at 1.55 μm and −31.5 dBm at 1.31 μm. The data show that the latest QD SOA generation easily enables trunk links with more than 60 km reach in combination with splitting ratios of 1:32.

Fig. 4.34(b) and Fig. 4.34(d) give insight into the dependance of the reach extension offered by the QD SOA as a function of the loss budget. For low losses before the QD SOA, the high input power causes gain saturation. Therefore the budget extension is small, but still exceeds 8 dB in the upstream case. If the losses before the devices are increased beyond 30 dB the OSNR of the data signal becomes too low, and consequently the budget extension decreases. The optimum loss budget extension is obtained for losses of around 20 dB before the amplifier. Fig. 4.34(b) and Fig. 4.34(d) further show that the 1.31 μm QD SOA offers a dynamic range of 40 dB. I.e., it tolerates input powers between +5 dBm and −35 dBm with BER < 10^{-9}. This dynamic range is enough to handle the total power fluctuations induced by the users in the network.

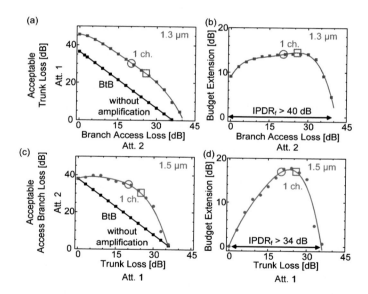

Fig. 4.34: Budget extension of QD SOA for the upstream and downstream path of a GPON. (a) Access branch loss plotted against acceptable trunk losses for the upstream path for which the BER is always better than 1×10^{-9}. (b) Budget extension of max. 14 dB by the QD SOA is achieved for losses of around 20 dB in front of the amplifier. The dynamic range (IPDR$_f$) is 40 dB. (c) Trunk loss plotted against acceptable access branch loss for the downstream path for which the BER is always better than 1×10^{-9}. (d) Budget extension of max. 17 dB by the QD SOA is achieved for losses of around 20 dB in front of the amplifier. The OLT and ONT launch powers are 5 dBm in both cases. The dynamic range (IPDR$_f$) is 34 dB, [C25].

To demonstrate the results of the budget extension analysis in a GPON, a testbed as in Fig. 4.32 is used. It mimics a situation with given losses in the trunk and the access branch; see large squares □ in Fig. 4.35. The setup comprises the CO with the OLT, a 50 km (SMF-28, see *Appendix A.5*) fiber trunk, the amplifier extender box with the QD SOA, the WDM de-/multiplexer followed by a 1:32 passive splitter, and the ONT. The OLTs in the CO are equipped with four WDM channel transmitters (1535 nm, 1545 nm, 1550 nm, 1560 nm) for the downstream data at 2.5 Gbit/s, and two WDM receivers (1290 nm and 1310 nm). Four ONTs are 10 km off from the extender box. The ONTs comprise four receivers, yet–due to lack of equipment–only two of the four ONTs are equipped with transmitters generating upstream traffic at 622 Mbit/s. The launch powers of the OLT and the ONT transmitters are +5 dBm. The receivers are identical with the ones used in the budget extension measurements above. The traffic for the present experiments is continuous. The loss budget in the 1.55 μm downstream link is 45 dB and 48 dB in the 1.31 μm upstream link. The polarization issue of QD SOA has been accounted for by using polarization controllers aligned to the maximum output power. In future field trials the most recent polarization insensitive QD SOA or a polarization diversity scheme as shown in [50] might be used.

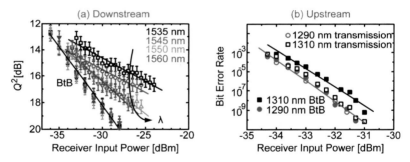

Fig. 4.35: (a) Downstream Q^2 factor and (b) upstream BER performance of the extended TDM/WDM GPON for all channels, [C25].

The open symbols in Fig. 4.35(a) represent the results of Q^2 factor measurements at 2.5 Gbit/s OOK for the four downstream channels in the 1500 nm band. Fig. 4.35(b) shows BER curves for the two 622 Mbit/s OOK upstream channels in the 1300 nm band (open symbols). Compared to the back-to-back (BtB) results (filled symbols), no power penalty is observed in the upstream case. However, a maximum power penalty of 5.5 dB is found for the downstream wavelength of 1535 nm at a $Q^2 = 15.6$ dB. This penalty is caused by saturation effects and channel cross-talk within the amplifier. No error floor and no cross-talk between upstream and downstream channels are observed even when the system is operated with all channels simultaneously.

4.6.2 Bursty On–Off–Keying Data Signals

In the upstream direction of a WDM/TDM GPON, the time division multiple access (TDMA) scheme in conjunction with different optical path lengths and/or different optical split ratios lead to a burst signal. Typical burst lengths in GPON networks are in the order of 100 µs. The packets of the n-th ONT have an average optical power level $P_{ONT,n}$. In the following, a burst mode transmitter is set up, which generates bursts of 100 µs length. The transmitter offers a burst ratio (BR$_{Tx}$) of 5 dB between two ONTs, which carries a 1.25 Gbit/s NRZ OOK data signal. The influence on the signal quality is studied for different average input power levels of the bursts with the SOA 7.

4.6.2.1 Amplification of a Bursty Data Signal with a QW SOA

The burst transmitter generates rectangular shaped clock signals at a frequency of 5 kHz (corresponding to 100 µs burst length, 20 ns rise and fall time). The burst length is in the order of the guard intervals usually placed between packets in a TDMA scheme of GPON. These bursts are modulated with a 1.25 Gbit/s NRZ OOK data signal with a PRBS7. The BR$_{Tx}$ at the transmitter is always 5 dB. The average optical input power level of the ONT 1 is adjusted to investigate the impact on the signal quality of a bursty TDM upstream data signal if amplified with a reach-extending SOA (SOA 7).

The results of the qualitative study are shown in Fig. 4.36. The burst signals at a wavelength of $\lambda = 1310$ nm are illustrated after the amplification with SOA 7. The recordings are made with a DCA triggered on the burst clock frequency. Within the bursts of ONT 1 and ONT 2 the 1.25 Gbit/s NRZ OOK data signal is present. The average input powers to the SOA of ONT 1 are varied (a) $P_{ONT1} = -20$ dBm, (b) $P_{ONT1} = -5$ dBm, (c) $P_{ONT1} = -1.5$ dBm. The optical input power of the burst signal determines the operating point of the SOA, and thus the gain compression.

For a weak GPON signal ($P_{ONT1} = -20$ dBm) the SOA is operated for both ONTs within the regime of almost constant gain, the so-called linear regime. The output burst signal is not disturbed and the burst ratio after the SOA (BR$_{Rx}$) is 5 dB. Thus the burst ratio is maintained.

In contrast, if the input power of the ONT 1 ($P_{ONT1} = -5$ dBm) is close to the saturation input power of the SOA (ONT 2 input power is 5 dB below) the 1.25 Gbit/s signal shows overshoots steaming from the 1.25 Gbit/s data-stream. As consequence of gain saturation, the burst ratio after the SOA (BR$_{Rx}$) is compressed to 3 dB.

If the input power is further increased ($P_{ONT1} = 1.5$ dBm), the overshooting of the 1.25 Gbit/s signal becomes more severe, and the burst ratio is further suppressed down to 1.5 dB. The overshoots within the 100 µs burst relaxes within a time scale of around 100 ps, which is approximately the time scale of the dominate SOA (SOA 7) dynamic.

In summary, the transmitted burst ratio (BR$_{Tx}$) of a GPON signal is reduced when at least one ONT exceeds the SOA saturation input power. The BR compression is an effect of the gain saturation, which is due to the fact that the gain of the high input power level ONT 1 is lower compared to the low input power level ONT 2.

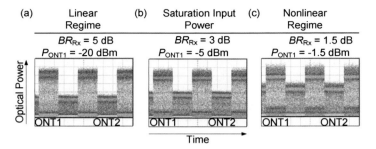

Fig. 4.36: Amplification of a bursty data signal with an SOA. (a, b, c): Burst signal (GPON comparable upstream data signal) at an optical wavelength $\lambda = 1310$ nm after amplification by the SOA 7 recorded with a DCA. The burst ratio at the transmitter (BR_{Tx}) between the ONT 1 and ONT 2 is always 5 dB in this study. The average optical power of ONT 1 P_{ONT1} is varied. If one of the ONTs has an input power level exceeding the saturation input power of the SOA overshoots of the 1.25 Gbit/s data signal occur. The burst ratio after the SOA (BR_{Rx}) is suppressed depending on the level of gain saturation.

4.6.3 IPDR of a SOA Cascade (ROADM) for OFDM and OOK Signals

SOA with a large IPDR are key devices to overcome losses in converged metro and access networks of the future. The losses compared to today's networks will dramatically increase due to a larger reach (easily exceeding 100 km), a higher split ratio (larger than 512) and the insertion loss of more network elements, see *Section 1.4*. One type of these network elements is a ROADM which can be used to enable versatile ring network infrastructures. To build metro and access ring networks which cover the area of a big city, ROADM need to be cascaded to permit a large number of subscribers, e.g., up to 100,000 [164]. These ROADM will incorporate loss-compensating optical amplifiers, e.g., SOA, which have to offer highest flexibility in coping with different multiplexing techniques, modulation formats, and optical channel power. Up to now, cascades of SOA have mostly been investigated for OOK data signals [139], [165], [166]. However, a study of SOA for metro and access ring networks for radio-signals such as IM-DD OFDM is still missing.

In this subsection, rules for the range of optical channel power of a cascade of SOA to guarantee the functionality of a metro and access ring network comprising several ROADM are demonstrated. Experiments in a recirculating fiber loop (17 km) with one QD SOA (SOA 2) are performed. The total propagation distances of up to 190 km employing IM-DD (electrical) OFDM-BPSK, OFDM-QPSK as well as OOK data formats are investigated. It is found that the NRZ OOK format is better in terms of achievable distance and input power dynamics compared to the OFDM-BPSK/QPSK formats. IM-DD OFDM can be applied especially in the downstream of such a network to serve a large number of antennas and mobile users.

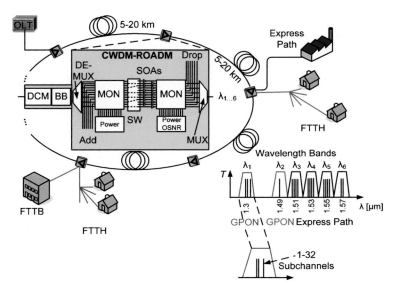

Fig. 4.37: Converged metro and access ring network scenario [164]. The loss compensating SOA are located in the ROADM which provide data switching every 5 km...20 km. This network supports infrastructures such as fiber-to-the-x (FTTx with x = B, C, H for building, curb, home) using 13 nm wide bands (20 nm separation) centered around $\lambda_{2...6} = 1.49$, $1.51,..., 1.57$ µm, and $\lambda_1 = 1.3$ µm. The GPON upstream and downstream channels are $\lambda_{1, 2}$. Express channels $\lambda_{3...6}$ are rent away to connect enterprises. Each of these bands contains up to 32 subchannels and is amplified by a dedicated SOA. The data signal on each subchannel is chosen flexibly. Therefore it is necessary to establish rules—dictated primarily by the SOA-for multiplexing techniques, modulation formats and channel power to guarantee a functional network.

Ring-Network Scenario:

The SOA investigated here are incorporated in a converged metro and access ring network scenario [164], Fig. 4.37, see also *Section 1.4.1*. Such a network offers virtually unlimited bandwidth for all kind of customers and infrastructures such as GPON and format-transparent "express paths" for business customers. Fig. 4.37 shows that from an OLT the data signal is added to the ring network through a ROADM. The data signals are transmitted over 5...20 km of fiber before they enter the subsequent ROADM. The ROADM comprise a dispersion compensating module (DCM), a black box (BB) unit, demultiplexer/multiplexer (DEMUX, MUX), and monitoring taps (MON) for power and OSNR monitoring. The BB unit equalizes the channel power levels within a wavelength band to avoid cross-talk, and it limits the maximum input power into the SOA to avoid pattern effects, phase distortions or intersymbol interference. The ROADM supports 6 CWDM bands ($\lambda_1...\lambda_6$) each separated by 20 nm and having a bandwidth of 13 nm for up to 32 subchannels. Two wavelength bands are reserved for upstream and downstream traffic in a GPON, and four wavelength bands are for the express paths. Each CWDM band is amplified by a dedicated SOA. The switch (SW) offers the flexibility to add and to drop each wavelength band, or to put it through to the next ROADM. Each subchannel can carry data employing different multiplexing techniques, modulation formats and channel powers. To guarantee a functional network, design rules for the subchannels must be provided which are primarily dictated by the SOA performance.

Experimental Setup:

The experimental setup (Fig. 4.38(a)) is a simplified realization of the scenario presented in Fig. 4.37. The aim is to investigate a subchannel performance as a function of the input power of the resulting SOA cascade and the possible span for OFDM and OOK data signals. A cw laser at a test wavelength of 1295 nm is gated with an SOA (Gate 1 time) = 72 µs, Fig. 4.38(b). The OFDM Tx uses intensity modulation to encode a real-valued electrical 5 GHz-wide OFDM signal generated by an arbitrary waveform generator onto an optical carrier. The OFDM signal consists of 58 modulated subcarriers (electrical subcarrier spacing 78.125 MHz) and 4 pilot tones centered around an intermediate frequency of 2.5 GHz. The total bit rate is 4.5 Gbit/s for BPSK modulation and 9 Gbit/s for QPSK modulation (including overhead), respectively. The OOK transmitter (Tx) generates 7 Gbit/s and 10 Gbit/s NRZ OOK signals with a pulse pattern generator. The loop comprises 17 km of SMF (loop propagation time 85 µs), a 3 dB coupler and the ROADM which is equipped with a CWDM filter (13 nm bandwidth), a gated SOA (device under test; (Gate 2 time) = 1 ms, Fig. 4.38(b)) and an attenuator (Att.) to mimic ROADM losses.

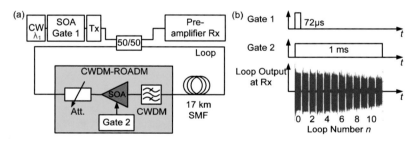

Fig. 4.38: Recirculating loop setup. (a) The loop comprises an SOA, an attenuator, a CWDM filter and 17 km of SMF fiber. OFDM-BPSK/QPSK and OOK signals are generated which are adjusted in power before launching them into the first SOA in the recirculating sequence. The receiver detects the signal passing *n*-times through the SOA. (b) Transmitter and QD SOA gate window and loop output signal at the Rx. A data signal at a test wavelength of 1.295 µm is chosen for the experiments, because the dispersion in the ring network scenario is always compensated with the DCM.

The SOA under test is the SOA 2 with a 1 dB saturation input power of −12 dBm. The QD SOA gain polarization dependance (10 dB) is accounted for by a polarizer and polarization controllers. The loop losses are adjusted once by the loop attenuator to compensate the QD SOA small-signal gain of around 17 dB. The OFDM and OOK output signals of the loop cycles are directly detected with a pre-amplifier Rx, respectively. The data signals are recorded by a real-time scope for offline processing and analysis. The Rx input power is kept constant well above its sensitivity threshold for a BER of 10^{-9}.

Experimental Results:

In Fig. 4.39 and Fig. 4.40 the achievable number of loop cycles for a range of channel input power entering the first SOA are presented considering different multiplexing and modulation formats. Fig. 4.39(a) indicates that for OFDM-BPSK 8 cycles are achievable with an EVM corresponding to a BER better than 10^{-3} (horizontal dashed line). In contrast, Fig. 4.39(b) indicates that only 4 loops are achievable with OFDM-QPSK modulation. In Fig. 4.39(c) the maximum and minimum input power levels (input power dynamic range given that BER $\leq 10^{-3}$) at the SOA input at the first loop cycle are presented as a function of the number of recirculating loop cycles (RL 1...11) (i.e., the number of cascaded SOA). The upper input power limit is given by nonlinear interaction of the OFDM subcarriers (inter-modulation) inside the SOA and by the modulation response of the SOA, see *Section 2.2.7*. The lower input power limit is set by the OSNR. The upper power limit monotonously decreases for increasing loop cycles. The lower power limit monotonously increases for increasing loop numbers for maintaining a sufficient OSNR. The optimal SOA input power lies around the 1 dB saturation input power (red dot).

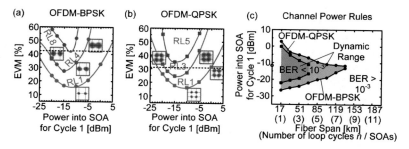

Fig. 4.39: EVM measurement for "cascaded" SOA in a recirculating loop (RL 1...11) transporting OFDM signals. EVM averaged over all subcarriers vs. input power entering the SOA at the first loop cycle for different loop cycle numbers and modulation formats. (a) 4.5 Gbit/s OFDM-BPSK, (b) 9 Gbit/s OFDM-QPSK with some constellations as insets. (c) Range of channel power for a choice of BER and spans. An increase in modulation format complexity from BPSK to QPSK significantly reduces the reach from around 136 km to around 68 km.

Fig. 4.40(a) shows that with a 7 Gbit/s OOK modulation 11 loop cycles are achievable with $Q^2 > 9.8$ dB corresponding to a BER $< 10^{-3}$. Fig. 4.40(b) shows that 10 loop cycles can be handled with a 10 Gbit/s OOK modulation. In Fig. 4.40(c) the dynamic range (given that BER $\leq 10^{-3}$) is presented. The upper input power limit occurs due to bit-pattern effects. The lower input power limit is due to a small OSNR.

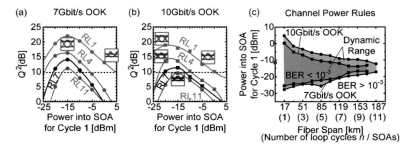

Fig. 4.40: Q^2 factor measurement for "cascaded" SOA in a recirculating loop (RL 1...11) transporting OOK signals. Q^2 vs. input power entering the SOA at the first loop cycle for different loop cycle numbers and bit rates. (a) 7 Gbit/s OOK, (b) 10 Gbit/s OOK with some eye diagrams as insets. The horizontal dashed lines show the Q^2 limit for a BER of 10^{-3}. (c) The OOK data signals show a significantly longer reach of 187 km and 170 km compared to the OFDM signals. The optimum input power into the SOA "chain" is near the 1 dB saturation input power (red dot).

Design Rules:

Based on the experimental results, the following rules (1...5) apply for cascaded SOA in a metro and access network:

(1) OOK modulation is advantageous compared to IM-DD OFDM with BPSK or QPSK modulation in terms of span number and input power dynamics.

(2) Lower-order modulation formats occupying the same bandwidth are preferable to achieve large span numbers. BPSK tolerates a higher EVM compared to QPSK (identical BER), and thus it has a lower OSNR requirement.

(3) The dynamic range reduces significantly for an increasing number of cascaded SOA, which limits the acceptable power variations that are introduced, e.g., in GPON upstream paths.

(4) The best input power to the first SOA lies near the 1 dB saturation input power, showing that linearity requirements are high.

(5) OFDM signals seem to be well suited for short-range connections [164].

Outlook:

From the discussion in this subsection and the results presented in the *Section 4.3* and *Section 4.4*, a prediction of measures which have to be taken to achieve an aggregated bit rate exceeding 1 Tbit/s in an "express paths" for the above scenario are provided.

In general, it is advantageous to use a low order modulation format with a large number of WDM channels with an SOA cascade. E.g., an express paths can be generated using data signals which are encoded with QPSK modulation (2 bits per symbol) on a 50 GHz spaced DWDM grid using 32 subchannels. If SOA devices with almost polarization independent gain are used, system operators could also think about the use of polarization-multiplexed DWDM-QPSK channels with a symbol rate of 25 GBd or even higher. A cascade of at least 5

SOA with this DWDM dual-polarization QPSK signals should be achievable in the near future. However, influences from phase-to-amplitude conversion during transmission and from four-wave-mixing in the SOA are expected to limit the cascadability. The use of higher order modulation format data signals than QPSK is very challenging. The use of 16QAM could be reasonable in the future if excellent advanced FEC, and ultra-linear SOA are used. The use of 32QAM or even higher M-ary QAM seems to be too challenging for amplification with an SOA cascade from today's point of view.

4.7 Conclusion and Comparison

In this chapter, the important parameters to build linear SOA are discussed. Device design guidelines for an SOA with a large IPDR are derived. In system experiments large IPDR values for such SOA are measured with OOK, PSK, QAM and RoF modulation format signals and with single channel and multi-wavelength channels. Possible applications of such linear SOA have been predicted and demonstrated. In the following, a summary and a comparison to results of other groups are given.

Linear Amplification of Intensity-Encoded Signals and Optimization of SOA Design Parameters for Large IPDR: The input power dynamic range of SOA has been discussed in general. A description relating device parameters with the IPDR is given. The model predicts large IPDR for SOA with a large modal cross-section, a low confinement factor and a long active region. These parameters are favorable especially for the upper IPDR limit by providing a large saturation power. Further a high bias current is preferable for a large IPDR. This is because a higher current provides a higher population inversion and thus lower ASE noise and it keeps the effective carrier lifetime low which helps to further increase the saturation power. Finally, to keep IPDR large the gain has to be chosen just to meet the need of the application. An overly large gain degrades the IPDR significantly.

These optimum specifications for a large IPDR favor QD SOA over other SOA types. This is because the number of QD-layers can be chosen to be small and a minimum gain can be maintained by choosing a long active region. Then, the IPDR description with QD SOA at 1.3 µm is experimentally validated. In experiments a large input power dynamic range exceeding 32 dB with Q^2 values of 12.6 dB for signals at bit rates of 2.5 Gbit/s, 10 Gbit/s and 40 Gbit/s in single channel operation is found. Even for operation with multiple signals (WDM) we still found an IPDR$_f$ in the range of 22 dB.

Finally, it should be emphasized that bulk, QW and QD SOA that come close or match the above derived design guidelines may as well be suitable to fulfill the requirements of next generation passive optical networks that require reach extender with IPDR$_f$ > 35 dB and FtF gain > 15 dB at a target BER of 10^{-9}.

Linear Amplification of Advanced Modulation Format Signals: The impact of the alpha factor on the amplification (IPDR) of signals with advanced optical modulation formats has been studied by simulation and verified by experiment. It is found that the influence of the alpha factor decreases with increasing complexity of the modulation format. This is due to a lower probability for large power transitions which in turn reduces the influence of gain changes and the associated phase errors. A low alpha-factor SOA is advantageous if it is used for purely phase-encoded signals (BPSK, (D)QPSK) under gain saturation. On the contrary, a larger alpha-factor SOA can also be successfully employed for M-ary QAM signals with a large number of different amplitude levels. In this case, a larger alpha-factor SOA does not deteriorate the signal more than an SOA with a low alpha factor.

Linear Amplification of Radio-over-Fiber Signals: Future RoF systems will be integrated into the fiber-based access networks. It is demonstrated that QD SOA reach extenders provide a large optical IPDR as well as a large EPDR. Optimum conditions are determined for QPSK and 16QAM radio-frequency signals. Experiments have been performed with 2 GHz and 5 GHz RoF signals.

Linear SOA Applications—Reach-Extender in Next-Generation Access Networks: The application of a QD SOA with a large IPDR as a reach extender for NG long-reach PON has been studied. NG long-reach PON requires 1.3 µm and 1.55 µm amplifiers for upstream and downstream, respectively. Burst-mode compatibility and largest possible input power dynamics are necessary. It is shown that the latest-generation QD SOA fulfill the demanding specifications. In-line amplifiers based on QD SOA enhance the total loss budget to 45 dB for a large range of input powers. An extended reach PON of 60 km with a split ratio of 1:32 with four downstream WDM channels at 2.5 Gbit/s OOK and with two WDM upstream channels at 622 Mbit/s OOK is demonstrated. It can be concluded that QD SOA are a promising technology for constructing PON-extender boxes if their polarization dependant gain is reduced in the future.

Linear SOA Applications—Cascade of QD SOA with IM-DD OFDM and Intensity-Encoded Signals: Fiber-loop experiments with one QD SOA show superior results for OOK signals in terms of achievable distance and IPDR compared to intensity modulated OFDM-BPSK/QPSK signals. Rules for the range of optical channel powers of a cascade of SOA to guarantee the functionality of a metro and access ring network comprising several ROADM have been extracted. IM-DD OFDM-BPSK (4.5 Gbit/s including overhead), IM-DD OFDM-QPSK (9 Gbit/s including overhead) data signals as well as 10 Gbit/s OOK data signals have been tested in a 17 km long fiber-loop.

Finally, the $IPDR_f$ values obtained in this thesis on QD SOA are compared to results of the $IPDR_f$ for other device technologies such as QW/bulk SOA, fiber-doped amplifiers, but also with schemes using filters, active gain clamping (AGC) or holding beams. It can be seen that the $IPDR_f$ values obtained on QD SOA in this thesis, without the use of any filtering or holding beam techniques, are comparable with the $IPDR_f$ values for other technologies which are using costly compensating techniques. Conventional QW/bulk SOA show lower $IPDR_f$ values.

The $IPDR_f$ results are presented in Table 4.8 and Table 4.9. The $IPDR_f$ is evaluated (Eval.) by introducing $IPDR_f$ category groups:

at a target BER of 10^{-9}: + 0...10 dB $IPDR_f$, ++ 10...25 dB $IPDR_f$, +++: > 25 dB $IPDR_f$,

at a target BER of 10^{-3}: o 0...10 dB $IPDR_f$, oo 10...30 dB $IPDR_f$, ooo > 30 dB $IPDR_f$.

The results at a target BER of 10^{-5} are treated along with the target BER of 10^{-3}.

OOK:

No.	SOA Type / Scheme	G_{ff} [dB]	$IPDR_f$ [dB]	Bit rate [Gbit/s]	Mod.	Multiplex	Target	Eval.	Ref.
1	QD	12	41	2.5	NRZ OOK	SC	BER = 10^{-5} (Q^2 measured)	ooo	[J2]
2	QD	16	28	2.5	NRZ OOK	WDM (2 ch)	BER = 10^{-5} (Q^2 measured)	oo	[J2]
3	QD	12	33	2.5	NRZ OOK	SC	BER = 10^{-9} (Q^2 measured)	+++	[J2]
4	QD	12	25	10	NRZ OOK	SC	BER = 10^{-9} (Q^2 measured)	+++	[J2]
5	QD	16	18	10	NRZ OOK	WDM (2 ch)	BER = 10^{-9} (Q^2 measured)	++	[J2]
6	QD	16	26	40	RZ OOK	SC	BER = 10^{-9} (Q^2 measured)	+++	[J2]
7	QD	16	20	40	RZ OOK	WDM (2 ch)	BER = 10^{-9} (Q^2 measured)	++	[J2]
8	AGC	0-28	40	10	OOK	WDM (4 ch)	BER = 10^{-9} (PP2 dB)	+++	[126]
9	LOA	15	> 25	10	OOK	SC	BER = 10^{-9}	+++	[125]
10	Holding	5	20	10	OOK	WDM (4 ch)	BER = 10^{-9} (PP2 dB)	++	[127]
11	Filter	18	32	10	OOK	SC	BER = 10^{-9} (PP 2 dB)	+++	[128]
12	PROF	n.a.	15	40	OOK	WC	BER = 10^{-9} (Q^2 measured)	++	[129]
13	EDFA	30	> 30	10	NRZ OOK	WDM (4 ch)	Q^2 = 15.6 dB (PP 2 dB)	+++	IPQ

Table 4.8: Comparison of $IPDR_f$ for OOK and for different technologies. $IPDR_f$ values No. 1-7 are measured within this work on a single SOA. IPDR values No. 8-12 are literature values obtained with sophisticated filtering or gain clamping schemes. $IPDR_f$ value for No. 13 is also obtained within the framework of this thesis on an EDFA. The type of the amplifier, the FtF gain, the $IPDR_f$, the aggregate bit rate, the modulation format, the multiplexing technique (SC: single carrier, WC: wavelength converted or WDM: wavelength-division-multiplexing), the target BER, and the evaluation of the $IPDR_f$ are shown.

In Table 4.8 the $IPDR_f$ results for the OOK modulation format are presented. The device type, the FtF gain, the $IPDR_f$ values, the bit rate, the modulation format, the multiplexing format including the number of channels in the experiments, and the target BER (including the evaluation) are presented. Obviously, the $IPDR_f$ results achieved in this thesis for OOK modulation can compete with technologies using costly filters, holding beams, sophisticated device structures or additional current supplies. The EDFA technology shows always the largest IPDR$_f$ values for OOK modulation formats if Gbit/s bit rates are applied. However, if such fiber-based amplifiers are operated in the upstream path of an access network they significantly suffer from overshoots due to the bursty nature of the traffic. Additionally, at a wavelength of 1.3 μm the PDFA technology is not a mass product and the devices are very expensive. Thus, especially in the upstream path of an access network a good designed QD SOA is a viable solution as a reach extender amplifier with OOK modulation.

Abbreviations in Table 4.8: LOA (linear optical amplifier), PROF (pulse reformatting optical filter), SC (single carrier), WDM (wavelength-division-multiplexing), AGC (active gain clamping), and EDFA (erbium-doped fiber amplifier).

(D)PSK / (D)QPSK / QAM / IM-OFDM:

No.	SOA Type	G_{ff} [dB]	IPDR$_f$ [dB]	Bit rate [Gbit/s]	Mod.	Multiplex	Target	Eval.	Ref.
14	QD	16	44	20	BPSK	SC	BER = 10^{-9} (EVM meas.)	+++	[J1]
15	QD	16	33	40	QPSK	SC	BER = 10^{-9} (EVM meas.)	+++	[J1]
16	QD	16	13	80	16QAM	SC	BER = 10^{-3} (EVM)	○○	[J1]
17	QD	16	32	56	DQPSK	SC	BER = 10^{-3} (PP2 dB)	○○○	[J6]
18	QD	16	36	56	DQPSK	WDM (2ch)	BER = 10^{-3} (PP2 dB)	○○○	[J6]
19	QD	17	30	5	BPSK	OFDM	BER = 10^{-3} (EVM meas.)	○○○	Ch. 4
20	QD	17	22	10	QPSK	IM-OFDM	BER = 10^{-3} (EVM meas.)	○○	Ch. 4
21	QD	17	> 30	0.02	QPSK	RoF SC	BER = 10^{-9} (EVM meas.)	+++	[C9]
22	Bulk	16	> 20	10	DPSK	WDM (8 ch.)	BER = 10^{-3} (PP 2 dB)	○○	[151]
23	Bulk	16	> 20	43	DPSK	WDM (8 ch.)	BER = 10^{-3} (PP 2 dB)	○○	[149]
24	Bulk	n.a.	< 15	10	RZ DPSK	SC	BER = 10^{-9} (PP 2 dB)	++	[153]
25	Bulk	n.a.	< 10	10	NRZ DPSK	SC	BER = 10^{-9} (PP 2 dB)	+	[150]
26	CC QW	12	> 10	40	RZ DQPSK	SC	BER = 10^{-3} (PP 2 dB)	○○	[154]

Table 4.9: Comparison of IPDR$_f$ for advanced modulation formats and for different device technologies. IPDR$_f$ values No. 14-21 are measured within this work on QD SOA. IPDR values No. 22-26 are literature values. The type of the SOA amplifier (current controlled: CC), the FtF gain, the IPDR$_f$, the aggregate bit rate, the modulation format, the multiplexing technique (SC, WDM, OFDM: orthogonal frequency-division-multiplexing), the target BER, and the evaluation of the IPDR$_f$ are shown.

For advanced modulation formats, it can be seen that the $IPDR_f$ values obtained with the QD SOA in this thesis are larger compared to other technologies. In Table 4.9 the $IPDR_f$ results for the advanced modulation format signals are presented. The device type, the FtF gain, the $IPDR_f$ values, the aggregate bit rate, the modulation format, the multiplexing format including the number of channels in the experiments, and the target BER (including the evaluation) are shown. Obviously, the $IPDR_f$ results presented in this thesis for advanced modulation formats as well as for IM-OFDM multiplexing techniques significantly exceed the $IPDR_f$ results obtained with, e.g., current controlled (CC) SOA or ordinarily designed bulk/QW SOA. Abbreviation in Table 4.9: IM-OFDM (intensity-modulated direct detected OFDM).

Generally, for SOA amplifying signals with bit rates of about 1...100 Gbit/s it is concluded:

- Low confinement factor, low noise figure, and low alpha-factor linear SOA, i.e., QD SOA, offer large IPDR values for different modulation formats (OOK, PSK, QAM, RoF) and multiplexing techniques (WDM, IM-OFDM).
- If moderate gain suppression cannot be avoided: An ultra-fast SOA gain recovery time (< 10 ps) is desirable for OOK applications, whereas a slow SOA phase recovery time (> 200 ps) is advantageous for PSK applications. Exactly this performance can be achieved with a single QD SOA.
- Best IPDR performance can be achieved with PSK/QPSK modulation formats followed from OOK signals. Higher-order QAM signals suffer significantly from SOA gain saturation.
- For QD SOA: Long devices with few QD-layers are required to optimize the IPDR for a given gain.
- The performed measurements with optimized QD SOA show larger $IPDR_f$ values compared to conventional QW/bulk SOA. Further, the measured $IPDR_f$ values on QD SOA are comparable to the $IPDR_f$ values found for holding beam, gain clamping or filtering techniques applied to conventional SOA.
- QD SOA are promising devices as in-line reach-extender amplifiers in next-generation access networks. Their large IPDR values enable their use in the upstream path of such networks (from customers to central office).
- QD SOA can also act efficiently as booster amplifiers if larger IPDR values with lower gains are chosen.
- SOA can be successfully concatenated for IM-OFDM and OOK data signals. More than 100 km fiber transmission can be realized. Thus, SOA are promising devices to be deployed as loss-compensating amplifiers in, e.g., ROADM.
- Well engineered bulk or QW SOA using the derived parameters for a linear SOA can offer comparable performance than its QD SOA counterpart.

5 Nonlinear SOA and Network Applications

Next-generation all-optical networks require a cost effective, low power consuming and small footprint realization of nonlinear signal processing devices. Here, conventional quantum well (QW)/bulk semiconductor optical amplifier (SOA) operated in the so-called nonlinear gain regime have already shown their potential to perform, e.g., wavelength conversion or clock recovery extraction at high speed.

In this chapter, the nonlinear operation regimes of novel quantum-dot (QD) SOA are investigated. SOA nonlinearities namely cross-gain modulation (XGM) and cross-phase modulation (XPM) are used to realize the application of a wavelength converter at different bit rates for on-off keying (OOK) data signals, and a clock recovery for 40 Gbit/s OOK signals.

In *Section 5.1*, fundamentals of wavelength converters and a clock recovery are presented. The content of the *Subsection 5.1.1* has been published by the author in [J7]. Minor changes have been done to adjust the notations of variables and figure positions.

In *Section 5.2*, desired SOA parameters enabling an efficient nonlinear element are provided. The parameters gain, noise figure, saturation input power, SOA dynamics, and the alpha factor are discussed.

In *Section 5.3*, the all-optical wavelength conversion using an ideal filter is derived theoretically, and the results are confirmed by experiments with 10 Gbit/s, 40 Gbit/s and 80 Gbit/s return-to-zero (RZ) OOK data signals. The content of the *Subsections 5.3.1, 5.3.2.1 and 5.3.3* has been published by the author in [J7]. Minor changes have been done to adjust the notations of variables and figure positions. Parts of the paper published by the author in [C28] are used to write the *Subsection 5.3.2.2*. Minor changes have been done to adjust the notations of variables and figure positions. The content of the *Subsection 5.3.2.3* has been published by the author in [J4]. Minor changes have been done to adjust the notations of variables and figure positions.

In *Section 5.4*, an experimental demonstration of an all-optical clock recovery at 40 Gbit/s RZ OOK using a Fabry-Pérot filter followed by a QD SOA is presented. High quality clock extraction with very low residual amplitude modulation and low root mean square (rms) timing jitter is achieved for single channel operation across the C-band (1530 nm-1565 nm). Two-channel clock extraction at 20 nm spacing illustrates the potential of the scheme for multi-wavelength operation. Part of the content of this section has been published by the author in [C14]. Several changes have been done to adjust the notations of variables, to the layout of figures and to the wording.

5.1 Fundamentals

In this section, fundamentals of wavelength converters based on XGM and XPM in SOA are presented. The wavelength conversion scheme consisting of an SOA followed by a filter is shown. The required filter as a function of the XGM and XPM properties of the SOA is discussed, and possible implementations are introduced. Further, fundamentals of a clock recovery circuit based on a Fabry-Pérot filter followed by an SOA are addressed.

5.1.1 Wavelength Conversion with SOA and State-of-the-Art

Wavelength converters are used to convert a data signal from a wavelength λ_1 onto a signal at another wavelength λ_2. The signal at the wavelength λ_2 can either be a continuous wave (cw) signal or a clock signal. Wavelength conversion requires a nonlinear interaction of the data signal with the signal at the wavelength λ_2. This interaction can be introduced by a nonlinear device, e.g., in SOA (*Section 2.2.8*), in a highly-nonlinear fiber or in especially designed waveguides based on Si (e.g., silicon-organic hybrid (SOH) slot waveguide) [129], [C24].

From a network point of view, wavelength converters are expected to play an important role in alleviating blocking issues in future high capacity transparent networks [168]-[170]. Among the various approaches, those based on bulk SOA have attracted considerable attention due to the small footprint of the devices and their low switching power requirements. In particular, the combination of a single SOA followed by an optical filter has demonstrated impressive performance at bit rates up to 320 Gbit/s OOK [68], [69], [72], [167], [171]-[176], [C38]. The success of this approach is attributed to the capability to overcome the speed limitations of the SOA. This is achieved by effectively mitigating the bit patterning distortions, which are introduced when the signal clock period exceeds the gain recovery time. The pattern-effect mitigation mechanism is fundamentally based on the correlation of phase and amplitude modulation, both of which are inherently attributed to the converted signal by the SOA. The optical filter then exploits this correlation, and through proper phase-to-amplitude conversion it counterbalances the patterning effect.

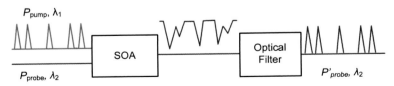

Fig. 5.1: Wavelength conversion scheme based on a nonlinear medium (SOA) and an optical filter. An inverted probe waveform results after the device. The optical filter restores the initial inversion and mitigates bit patterning, [J7].

Recently, self-assembled QD SOA have appeared showing the potential to provide significant performance benefits over bulk semiconductors. Due to their atom-like density of states and the existence of a carrier reservoir, those devices promise more efficient carrier refilling that may enable ultra-fast performance [116], [177], [178]. However, QD SOA are also known for their low alpha factor, which eliminates XPM as the dominant nonlinear switching mechanism and leaves XGM as the important effect, see *Section 2.3*.

Wavelength conversion based on XGM in QD SOA has been demonstrated at various bit rates [50], [179]. However, similarly to bulk SOA, XGM in QD SOA is characterized by a poor extinction ratio, which sets severe cascadability limitations. On the other hand, four-wave-mixing (FWM) based approaches have demonstrated tunability over several THz [179], [180] but they have a limited efficiency factor. Filter assisted wavelength conversion remains a challenging area for QD SOA. Although some first experiments have been already presented [181], [182], it is still unclear which should be the optimum schemes that taking into consideration the particular properties of QD devices can ensure a high quality operation similar to bulk SOA.

A generic configuration of a wavelength converter composed of an SOA and an optical filter is illustrated in Fig. 5.1. The optical filter might be a cascaded structure of optical band-pass filters (OBPF) or delay interferometers (DI). A strong data signal at wavelength λ_1 (pump) modulates the phase and amplitude of a cw signal (probe) entering the SOA at a different wavelength λ_2. At the output of the SOA this results in an inverted pulse-stream, which carries the information of the input data signal. Both pump and probe wavelengths are then directed to the optical filter, which has three functions. Firstly, it blocks the pump channel, secondly it restores the non-inverted waveform of the probe channel, and thirdly it suppresses bit patterning. The optimum transfer function depends strongly on the strength of XPM and XGM that the signal has experienced inside the SOA device. Theoretical and experimental studies show that this is primarily an issue of the SOA's operating conditions [116].

In the following, the focus is on wavelength conversion using QD SOA. The role of the optical filter differs for each operating regime of the QD SOA (*Section 2.3*). In case of low carrier injection (see *Section 2.3.2*), the filter, apart from inverting the signal waveform, has also to suppress the bit pattern distortions introduced by the dominance of the slow wetting layer (WL). In the high carrier injection regime, since pattern effects do not exist, the role of the filter is only to invert the signal waveform and to increase the extinction ratio of the converted signal.

For identifying the optimum filter response, a general solution is given in [129]. The transfer function $H_{\text{PROF}}(\omega_s)$ of such a filter, also called a pulse reformatting optical filter (PROF), is obtained by dividing the spectrum of the desired optical signal $E_{\text{desired signal}}(\omega_s)$ by the optical signal at the output of the SOA $E_{\text{out,SOA}}(\omega_s)$, for a fixed operating point (OP) of the SOA:

$$H_{\text{PROF}}(\omega_s) = \left. \frac{E_{\text{desired signal}}(\omega_s)}{E_{\text{out,SOA}}(\omega_s)} \right|_{\text{OP} = \text{const}} . \tag{5.1}$$

The result depends strongly on the OP of the QD SOA. In the low carrier injection regime, see *Section 2.3.2*, the optimum filter splits off the blue and the red signal components, equalizes their amplitudes, see Fig. 5.2(a), compensates their relative delay difference, see Fig. 5.2(b) [129], and finally recombines them. This process provides also an inherent mechanism for compensating pattern effects [82].

In the regime of high carrier injection, see *Section 2.3.2*, phase effects do not exist, and the converted signal is totally unchirped, which results in a symmetrical spectrum. Therefore, there is no need to introduce any relative attenuation or delay between the red and blue spectral components. A notch filter that through suppression of the carrier frequency achieves an inversion of the signal waveform may be considered an optimum solution for the present case. Fig. 5.2(c) and Fig. 5.2(d) depict the amplitude and the group delay response of such a filter type, respectively.

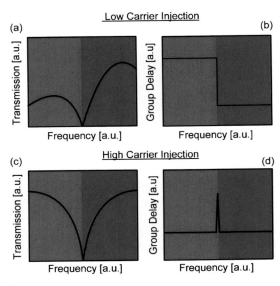

Fig. 5.2: Transmission and group delay of the ideal filter for the different operating regimes (see *Section 2.3.2*) of a QD SOA. In the low carrier injection regime the filter needs to equalize (a) the amplitude levels of the blue and red chirped spectral components as well as to (b) compensate their relative group delay difference. In the high carrier injection regime the optimum solution is a notch filter that introduces (c) no relative attenuation or (d) group delay difference between the red and blue spectral frequencies, [J7].

A PROF structure is generally complex and difficult to implement. It requires slicing of the signal's spectrum, introducing well defined time delays for the different spectral components, controlling their relative phases, and recombining the spectral constituents with appropriate attenuations at the output [129]. Therefore, other simpler schemes have been proposed performing single sideband selection (blue or red) of the signal spectrum [69], [171]-[176]. Operation at high bit rates can be ensured also for this case by an appropriate pattern-effect mitigation mechanism that occurs at the rising slopes of the frequency transfer function. The exact frequency position and the magnitude of the filter's slope are defined by the XPM and XGM properties of the device [173]. Given the weak XPM performance of QD SOA it needs to be investigated whether the mitigation of the bit patterning can be achieved efficiently with a more practical filtering scheme. Here, appropriate filters to support wavelength conversion at different operating regimes of the QD device are identified.

For the case of low carrier injection, the structure of two cascaded DI and an OBPF is proposed, Fig. 5.3(a). The filtering scheme has a periodic transfer function that can selectively suppress different regions of the signal's spectrum, Fig. 5.3(b). Therefore, by proper tuning of the phases and the delays of the two DI the frequency carrier of the converted waveform can be suppressed and meet the required slope matching conditions for effectively mitigating its bit patterning [183]. In addition, by adjusting the offset frequency of the OBPF, the scheme selects either the red or the blue sideband of the signal, Fig. 5.3(b). Comparing to other similar approaches, i.e., that of the combination DI+OBPF, the configuration proposed here is more flexible and offers higher adjustability to various power levels and device-specific properties.

For the case of high carrier injection, and since pattern effects do not exist, only a filter that inverts the signal waveform and improves its extinction ratio is needed. Therefore, a scheme with a single DI, Fig. 5.3(c), is proposed. The DI acts as a notch filter that suppresses the carrier of the signal and selects both the red and the blue spectral components with equal strengths, Fig. 5.3(d). Therefore, for the specific operating regime, the single DI can be considered as an easy and straightforward implementation of the PROF filter.

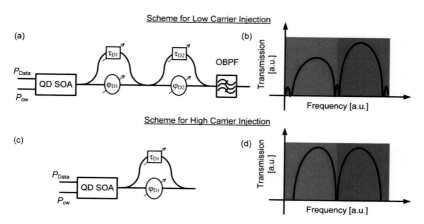

Fig. 5.3: Proposed filtering schemes for the different operating regimes of the QD SOA. For the low carrier injection regime, (a) a structure consisting of two cascaded DI and an OBPF is proposed. (b) By proper tuning its elements either the blue or the red sideband of the signal can be selected. In the regime of high carrier injection, (c) a scheme with a single DI is proposed. (d) It acts as a notch filter and transmits both red and blue spectral components alike, [J7].

Simulation results for the low carrier injection regime are presented in *Section 5.3.1.1* and the experimental realization at 40 Gbit/s is shown in *Section 5.3.2.1*. The simulation results for the high carrier injection regime are presented in Section 5.3.1.2 and a model experiment is demonstrated in Section 5.3.2.2. Results for a high-speed wavelength conversion experiment performed with a QD SOA under moderate carrier injection at 80 Gbit/s are demonstrated in *Section 5.3.2.3*.

5.1.2 Optical Clock Recovery and State-of-the-Art

In a communication system, a data-stream is transmitted across a medium (fiber, air, copper cable) from a transmitter (Tx) to a receiver (Rx) with no accompanying timing reference due to the typical large spatial separation between Tx and Rx. Clock recovery units (CRU) are used to recover the timing information (data clock) from a data signal. The recovery of the clock from a stochastic data signal (e.g., a pseudo-random binary (bit) sequence: PRBS) requires a spectral component at the clock frequency. For RZ modulated data signals this requirement is achieved, because there is a corresponding power spectral density at the clock frequency. However, for NRZ modulated data signals a frequency component at the clock frequency needs to be generated by a nonlinear operation.

An incoming signal offering a spectral component at the clock frequency is used to synchronize with an internal oscillator within the CRU. This means that the incoming signal acts as an external reference signal, and the CRU obtains information about the frequency and the phase of this reference signal. The clock signal recovered by the CRU has to respond very slowly in frequency and phase on changes of the incoming data signal. Thus, the CRU should offer high resilience to the timing jitter of the incoming data signal. Further, the data signal may exhibit long sequences of "1" or "0" requiring that the CRU still locks to the clock signal.

At some positions in the network an optical-to-electrical conversion (OEO) of an optical data signal is unnecessary. Thus, an optical CRU has to be implemented by preserving the signal in the optical domain. Optical CRU are based on active pulsating techniques or passive filtering techniques. The discussion follows closely the review of [184]. A common self-pulsating technique for the implementation of an optical CRU is based on an optical phase-locked loop (PLL), which offers closely the functionality of an electrical PLL. This technique has been successfully used for the clock recovery of optical time-division-multiplexing (OTDM) data signals at bit rates of up to 320 Gbit/s and 640 Gbit/s [185], [186].

Self-pulsating lasing is another technique to build an optical CRU. The frequency of the laser optical output clock signal can be locked to the reference frequency of the optical input signal. This technique has been successfully used to recover the clock from up to 160 Gbit/s OTDM data signals. Using the active pulsating technique high-quality clock signals can be generated, however, usually at the price of high device complexity and manufacturing costs.

On the contrary, a CRU does not necessarily require self-pulsating techniques. Passive filtering methods can be used to recover the clock. The passive filtering technique, e.g., Fabry-Pérot (FP) filter or Bragg grating filter [187], typically has a simple construction and low manufacturing cost, however, the challenge is to produce a high-quality clock signal. A clock signal can be recovered from a data signal with a FP filter if its free-spectral range (FSR) matches the bit rate for an incoming RZ OOK data signal. The issue of the FP filter based CRU is that the output clock pulses can significantly fluctuate in power (bit-pattern effects) depending on the resonator bandwidth and the input pattern of the data signal. To perform a clock recovery for as many consecutive low-level bits ("0") as possible and to minimize the bit-pattern effect, a FP resonator with the lowest possible bandwidth is preferred. An electric field that decays as $\exp(- t/2\,\tau_{ph})$, corresponding to an energy that decays as $\exp(- t/\tau_{ph})$, has a Fourier transform that is proportional to $(1/(1 + j\,4\pi f\,\tau_{ph}))$, which has a FWHM spectral width of $\Delta v_{FP} = 1/(2\pi\,\tau_{ph})$. Thus, the photon lifetime τ_{ph} is inversely proportional to the FP filter bandwidth Δv_{FP}. The number of resolved consecutive "0" level bits are determined by the photon lifetime. E.g., 10 (30) consecutive "0" level bits are resolvable at a bit rate of 40 Gbit/s with a FP filter bandwidth of around 700 MHz (about 200 MHz). The resulting bit-pattern effects from the energy decay due to the long sequence of "0" level bits can be mitigated with an SOA which is operated in the nonlinear gain regime. FP filters and Bragg grating filter have been used to recover the clock from data signals at bit rates of up to 40 Gbit/s with OOK modulation [188]-[194].

The functionality of an optical multi-wavelength clock recovery have attracted much attention due to the advantage of treating multiple channels independently, thereby reducing size, cost and power consumption [195]. Multi-wavelength clock recovery has been demonstrated so far for 10 Gbit/s traffic exploiting Brillouin scattering and Talbot effect in fiber-based configurations [196], [197] or using an integrated SOA array module for WDM channels [198]. Recently, a clock recovery for 35.811 Gbit/s RZ data signals using a FP-SOA has been shown for two channels. However, the polarization state of each data signal into the SOA was always set to the best recovered clock performance ignoring cross-talk effects stemming from XGM and XPM in the SOA [199]. Therefore, the main limitations of SOA-based multi-wavelength operation are neglected in the aforementioned study. Finally, multi-channel clock recovery operation for optical packets has only been investigated numerically at 40 Gbit/s and 160 Gbit/s using a QD SOA-based FP circuit [200].

In this work, an optical CRU for RZ OOK data signals based on a FP filter and followed by a QD SOA will be introduced and experimentally investigated, see *Section 5.4*. The multi-wavelength capability of the proposed scheme will be evaluated.

5.2 SOA Parameter for Nonlinear Applications

In this section the influence of SOA parameters and their relevance for optical networks is discussed. In order to get highest nonlinear signal processing performance, parameters such as gain, noise figure, saturation power, SOA dynamics as well as the alpha-factor need to be optimized.

Important Parameters of Nonlinear SOA for, e.g., OOK Wavelength Conversion

The SOA *unsaturated FtF gain* G_{f0} should be in the range of 5 dB...10 dB. Ordinarily, a wavelength converter consists of an SOA followed by a filter. At least a small XGM (extinction ratio: ER around 2...3 dB) after the SOA is required so that the filter can improve the ER of the signal. However, the FtF gain should not be too large avoiding strong influences from bit-pattern effects.

The *FtF noise figure* (NF$_{ff}$) is of minor interest as long as the output signal of the SOA has an acceptable optical-signal-to-noise ratio (OSNR). NF$_{ff}$ values of 8...10 dB are acceptable. If the SOA should be used in a regenerative wavelength conversion scheme, the importance of the NF$_{ff}$ increases.

The in-fiber *saturation input power* $P_{sat}^{in,f}$ should be moderate, and in the range of $-15...-5$ dBm. This way saturation of the amplifier can be easily achieved at moderate input powers of the data signal.

The *90 %-10 % gain recovery time* has to be small to avoid bit-pattern effects. Since wavelength converters in future photonic networks require to operate at speeds beyond 100 Gbit/s OOK the SOA dynamic needs to be able to follow this data signal.

Lastly, a large *alpha factor* in the range of 5...8 has advantages as well. It guarantees that there are large phase variations upon a change in the gain. At least a small XPM (phase change of about $0.2\,\pi$) after the SOA is required so that the filter can mitigate bit-pattern effects in the amplitude by phase-to-amplitude conversion.

Influence of Dimensionality of the Electronic System on the Nonlinearity of an SOA

With respect to their suitability for a nonlinear amplifier, we now compare the SOA parameters for the mature bulk SOA and QW SOA along with the QD SOA. This is done from a theoretical point of view as based on the equations given in the *Section 2*, and the discussion of the linear SOA parameters in the *Section 4.1*, but also relying on experimental studies performed in recent years [J7], [J4], [C28].

The required *FtF gain* G_{ff} of 5...10 dB can be achieved with both bulk and QD SOA. Following (4.1), a moderate FtF gain can be achieved in bulk SOA with a large optical confinement factor Γ and a short device length L.

The *FtF noise figure* NF$_{ff}$ of 8...10 dB can be achieved with both bulk and QD SOA.

The *saturation input power* should show moderate values. Therefore, the modal cross-section C/Γ is selected to be moderate and the differential gain a is chosen to be large. The parameters C/Γ and a are chosen to counterbalance the influence of the gain on the evolution of P_{sat}^{in}. A large optical confinement factor causes the term C/Γ to be moderate (assuming a fixed C).

Technologically speaking, a large optical confinement factor is much easier achievable with bulk SOA compared to QD SOA. The differential gain a is comparable in current devices.

The 90 %-10 % gain recovery time may be smaller in QD SOA if they are operated in the high carrier injection regime (see *Chapter 2*). However, due to the fact that in nonlinear applications the SOA is operated under gain suppression, a large signal photon density S may significantly reduce the total effective carrier lifetime even in conventional SOA. Additionally, a high current density decreases the effective carrier lifetime in both bulk and QD SOA.

The phase change according to (2.7) is larger in bulk SOA due to the larger alpha factor α_H and the larger confinement factor Γ in these devices.

It can be concluded that in general the use of conventional SOA are advantageous for nonlinear applications. However, since QD SOA can show a very fast gain recovery it is interesting to investigate their suitability for nonlinear signal processing. Therefore, in this work, the relatively new QD SOA technology is studied experimentally for wavelength conversion and clock recovery at highest speeds with an OOK modulation format.

5.3 Wavelength Conversion Schemes and Its Realization

In this section the optimum filter designs for wavelength conversion in the regimes of low and high carrier injection is investigated. The analysis includes theoretical performance evaluation and experimental implementation of the proposed schemes. Finally, design rules for ultra-fast nonlinear QD SOA achieving pattern-effect free operation in the high carrier injection regime are presented, and their wavelength conversion performance is investigated.

5.3.1 Theoretical Investigation of Optimum Filtering Schemes

Equations (2.57)-(2.62) of the numerical QD SOA model from *Section 2.3.3* represent an approach with physical insight to analyze the XGM and XPM properties of the QD SOA. Yet, in order to accurately predict the performance of a device, the relevant parameters need to be determined by fitting measurement data from an actual QD SOA sample (SOA 5) to the QD model. The obtained parameters are listed in Table 5.1. The geometrical parameters (length, width, height and number of QD layers), correspond to the QD SOA sample (SOA 5), which is used to perform the experiments. The values of the remaining parameters are identified through the fitting process.

Fig. 5.4(a) demonstrates the fitting, which has been achieved for the steady-state FtF gain and the FtF noise figure performance versus the in-fiber SOA input power. The QD SOA sample (SOA 5) is characterized at the pump wavelength $\lambda_1 = 1550$ nm by an unsaturated FtF gain of 18 dB and an in-fiber saturation input power of -12 dBm. The FtF noise figure of the device is 9 dB. It is clearly to be seen that the results provided by the theoretical model are remarkably close to the measurements. The experimental characterization of the phase dynamics of the device along with the theoretical fitting is presented in Fig. 5.4(b). The measurements have been done with a 40 Gbit/s sequence of '1010...' launched into the sample together with a cw signal. An electro-absorption gating technique based on spectrograms was used to evaluate the corresponding phase changes of the cw signal [123]. Large input powers introduce stronger depletion in the wetting layer and thus larger peak-to-peak (PtP) phase changes. The measured modulus of the PtP phase changes as a function of the average in-fiber input pump power in Fig. 5.4(b) originate from the depletion of the WL. This indicates that although the device is operated at its driving current limit, complete filling of the QD states from the WL has been not yet achieved.

Symbol	Parameter	Value
J	Current density	12 kA/cm^2
η_i	Injection efficiency factor	0.3
N_D	Area density of dots	$0.79 \times 10^{15} \text{ m}^{-2}$
l	Number of layers	6
L	Amplifier length	2 mm
t_{wg}	Amplifier thickness	0.15 μm
w	Amplifier width	1.25 μm
τ_c	Carrier lifetime	100 ps
τ_1	Characteristic relaxation time	3 ps
$\hbar\gamma_{hom}$	Homogeneous linewidth	12 meV
$\hbar\gamma_{inhom}$	Inhomogeneous linewidth	40 meV
σ_{res}	Resonant cross-section	$1.3 \times 10^{-19} \text{ m}^2$
α_{int}	Waveguide losses	500 m^{-1}
α_H	Henry factor	2.4
N_{wL}	WL area density of states	$1.08 \times 10^{16} \text{ m}^{-2}$
W_{bind}	Average binding energy	150 meV

Table 5.1: Physical parameter values for the QD SOA (SOA 5), [J7].

To theoretically characterize the achieved QD occupation probability the corresponding QD filling factor averaged over the length of the device $\bar{W}_0(N) = (1/L)\int_0^L W_0(N,z)\,dz$ is considered. This quantity depends on the instantaneous value of the carrier density N and is minimized when the device is saturated by a long sequence of consecutive "1". If even in this case we still get $\bar{W}_0(N) = 1$ then the QD device is considered to operate in the high carrier injection regime. Such a simulation approach has been performed. A 256-bit PRBS datastream consisting of 8 ps Gaussian pulses repeated at 40 Gbit/s having an average in-fiber input pump power of 0 dBm is launched to the SOA. The stream contains the data pattern "011111", which may temporarily saturate the device at the highest level. The average in-fiber power of the probe channel is 0 dBm, too. The other device parameters are according to Table 5.1.

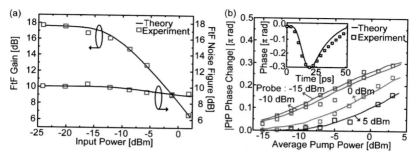

Fig. 5.4: Comparison of model and experiment. (a) Steady-state FtF gain and FtF noise figure characterization of the QD SOA sample (SOA 5) used in the experiment and fitting with the theoretical QD model. (b) Characterization of the dynamic modulus peak-to-peak (PtP) phase changes for different in-fiber power levels of the '1010' pump and cw (probe) signals. The solid lines result from the model and are fitted to the experimental data (□), [J7].

Fig. 5.5 illustrates the temporal minimum $\overline{W}_{0,min}$ of the QD filling factor as a function of the injected current density J. The maximum drive current density for the sample is 12 kA/cm². For these conditions, $\overline{W}_{0,min}$ amounts only to 0.53. The model estimates complete QD filling when the current density approaches 60 kA/cm². Of course, this is a theoretical limit far beyond the current density which the device (SOA 5) tolerates. In the following *Section 5.3.3* alternative designs for QD SOA are investigated allowing complete QD filling at much lower current densities.

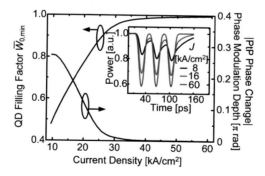

Fig. 5.5: Minimum filling factor of the QD states of the QD SOA as a function of the injected current density. Complete population inversion between the QD and the WL is theoretically achieved for a current density of 60 kA/cm². When increasing the injection current density, the phase modulation depth (modulus of PtP phase change) of the inverted pulses-stream decreases. Inset: Inverted signal waveforms for different values of the current density. Similar as in the previous cases, higher current densities lead to pattern-effect free operation, [J7].

The inset of Fig. 5.5 illustrates inverted data patterns for different current densities. It is clearly to be seen that as one moves towards the higher injection current regime, bit patterning is reduced. The same occurs for the phase effects. In Fig. 5.5 the minimum phase modulation depth of the pulse-stream is also shown. For a complete QD filling at 60 kA/cm^2, depletion is attributed only to the QD states, a fact which leads to zero phase changes.

5.3.1.1 Low Injection Current Regime – Simulations

Fig. 5.6(a) illustrates the proposed configuration that consists of two cascaded DI and an OBPF. In this structure each element has a distinct role to play. The first DI is mainly used to suppress the carrier frequency and thus to re-invert the wavelength converted signal. The second DI provides the pattern-effect mitigation. Finally, the OBPF performs the single sideband selection. The corresponding transfer functions are illustrated in Fig. 5.6(b). Free tuning parameters are the delay and the phase of each DI as well as the central frequency of the OBPF. The reciprocal of the delay $1 / \tau_{Dk}$ defines the frequency difference between two consecutive minima of the transfer function of the DI, while the phase φ_{Dk} defines their exact spectral position. In this case, each phase φ_{Dk} ranges within $[-\pi, \pi]$. For $\varphi_{Dk} = 0$ the DI suppresses the carrier frequency of the signal. For $\varphi_{Dk} > 0$ the transfer function minimum is on the red side of the signal spectrum, whereas for $\varphi_{Dk} < 0$ it is on the blue side. The OBPF has a Lorentzian line shape and a 3 dB bandwidth of 1 nm. In the experimental realization of this scheme, both DI are free-space optical setups. However, the same structure has been realized by integrating the DI in silica waveguides, and is known as 'optical equalizer' [201].

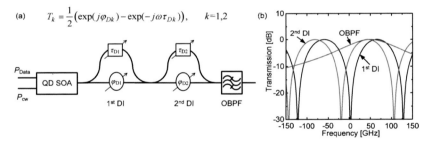

Fig. 5.6: (a) Proposed filtering structure of two cascaded DI and an OBPF. Free parameters are the phase and the delay of each DI as well as the offset frequency of the OBPF (Lorentzian). (b) Transfer functions of the corresponding filtering elements, [J7].

The exact transfer function of the proposed scheme is identified through the following optimization process: At the input of the device a data channel with an average in-fiber power of -5 dBm and a cw channel with an in-fiber power of 0 dBm are considered. As in Fig. 5.5, the data signal is a 256-bit PRBS stream of 8 ps wide Gaussian pulses with a bit-period of 40 Gbit/s. The input waveform is not ideal, but perturbed with some initial noise, that constitutes a realistic Q^2 factor value of 22 dB. The delays of the two DI are kept constant and equal to 8 ps, and the quality of the output signal as a function of the DI phase shifts φ_{D1} and φ_{D2} are evaluated. The corresponding results are illustrated in terms of contour plot diagrams in Fig. 5.7(a…c). Fig. 5.7(a) depicts the amplitude variation of the output '1's, defined by the ratio of the standard deviation σ_1 of the peak power levels to their mean value μ_{m1}. Although ASE noise degradation also influences the converted waveform, the variation on the peak power is dominated in this case by the bit-patterning effect. Therefore, lower values of the σ_1 / μ_{m1} rate indicate a better suppression of this phenomenon. This occurs when the transfer function minimum of each DI is away from the carrier frequency, thereby attenuating the blue ($\varphi_{D1} = -0.2\,\pi$) or the red ($\varphi_{D2} = 0.3\,\pi$) spectral regions of the signal, respectively. According to Fig. 5.7(b) this leads to severe extinction ratio (ER) degradation (ER $=\mu_{m1} / \mu_{m0}$, with the mean value of the output '0's' μ_{m0}).

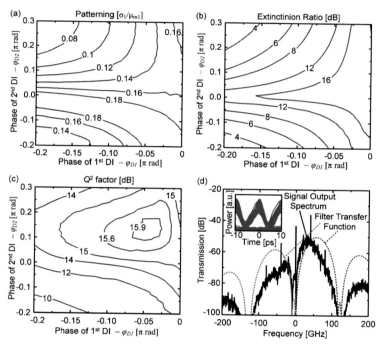

Fig. 5.7: Performance evaluation of the proposed scheme in the regime of low carrier injection. (a) Bit patterning induced amplitude variation of the output '1's', defined by the ratio of the standard deviation of the peak power levels to their mean value, (b) extinction ratio and (c) Q^2 factor of the signal at the output of the wavelength converter as a function of the introduced phase shifts φ_{D1} and φ_{D2} by the corresponding DI. (d) Spectrum and eye-diagram (inset) of the output signal and overall filter transfer function at the point of maximum Q^2 factor, [J7].

To identify a compromise between the two extreme signal quality cases, the Q^2 factor as the most appropriate figure of merit is considered. Fig. 5.7(c) depicts the results of the corresponding calculations. An optimum phase combination is noticed that gives a maximum Q^2 factor of 15.9 dB. In the previous cases the frequency offset of the band-pass filter has been optimized at 45 GHz. At the highest Q^2 factor point the output signal spectrum and the overall filter transfer function, see Fig. 5.7(d), have been calculated. The inset shows the corresponding eye diagram. The filtering scheme selects the blue sideband of the signal spectrum and discards the red one. Furthermore, it suppresses the carrier and uses the first modulated tone of the probe channel to form a new RZ spectrum centered at the offset frequency of 40 GHz. Finally, the rising edge of the filter's frequency response, at a small frequency offset from the carrier, is responsible for the pattern-effect mitigation mechanism.

The process of suppressing the bit patterning distortions is explained in Fig. 5.8(a)-(e). An exemplary input data signal with a bit pattern '11110' is launched into the QD SOA and gradually saturates its gain, see Fig. 5.8(a). Due to the Kramers-Kronig relation this gain change is always accompanied by a phase variation, which determines an instantaneous frequency shift $\Delta f_s = (1/2\pi) (\Delta\phi)/(\Delta t)$ (chirp). The leading edges of the input pulses induce a red chirp, while the trailing edges induce a blue chirp, Fig. 5.8(b). The red chirp takes on a larger value for the first '1' bit in the pulse train due to the stronger carrier depletion and the gain reduction that it induces. However, for subsequent pulses, as the gain reduction due to the carrier depletion decreases, the induced red chirp decreases as well.

For the blue chirp the situation is quite different. As the saturation level of the SOA is increased by each subsequent pulse, the carrier recovery becomes faster leading to a stronger blue-chirp [82].

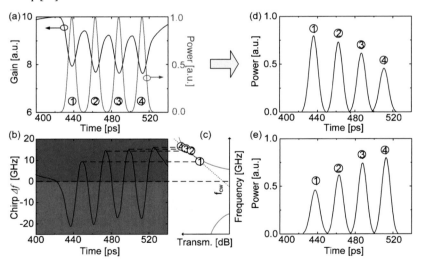

Fig. 5.8: Explanation of pattern-effect mitigation. (a) Gain evolution in the QD SOA induced by an input data pulse train and (b) induced frequency chirp in the inverted signal. (c) Schematic shape of the optimum transfer function of the filter. Generated signal waveforms from (d) AM/AM and (e) PM/AM conversion with opposite patterning behavior, [J7].

We may consider the influence of the overall filter decomposed into two separate mechanisms, which act on the amplitude and phase waveforms of the converted signal. In the first case we have an amplitude-to-amplitude (AM/AM) conversion mechanism, which is responsible mainly for re-inverting the signal. Due to the increasing gain saturation this mechanism creates a bit pattern having a decreasing average power level, as shown in Fig. 5.8(d). Apart from AM/AM conversion, phase-to-amplitude conversion (PM/AM) takes place, too, and it is illustrated in Fig. 5.8(b),(c),(e). In case of Fig. 5.8(c) the transfer function of the overall filter will suppress the red-chirped components and allows only the blue ones to pass through. In addition, the strongest blue chirped pulses will be favored from the filter's response, leading to an increasing power level for consecutive '1's, as shown in Fig. 5.8(e).

The combination of the AM/AM and PM/AM generated waveforms gives the total output signal. Their complementary pattern behavior indicates that the patterning can be mitigated as long as the phase-to-amplitude conversion at the slope of the filter compensates the saturation of the gain.

5.3.1.2 High Injection Current Regime – Simulations

When increasing the current density the region of high carrier injection is entered, where bit patterning induced by the QD SOA does not exist. In this case a notch filter is required to suppress the central frequency of the probe signal thus re-inverting the signal waveform. The single DI with a phase of $\varphi_D = 0$ can be considered as a possible implementation of such a filter type.

Fig. 5.9(a),(b) illustrates the operating principle of this DI scheme. The wavelength-converted signal (probe) enters the interferometer. There it is split and experiences a differential delay of τ_D. If the optical fields in the lower (point-A) and upper (point-B) arm of the DI have the same amplitude, a '0' field results at the destructive (difference) Δ-port, otherwise, a non-zero field strength appears. Careful adjustment of the delay parameter τ_D is required. For $\tau_D = 8$ ps, the DI generates two pulses within the same bit slot, see Fig. 5.9(c). The pulses have identical envelopes but differ in phase by π. Such a signal waveform is not acceptable for transmission.

The solution to this problem is depicted in Fig. 5.9(c). By choosing a differential delay of 1 bit ($\tau_D = 25$ ps), we avoid duplication of the generated pulse-stream. The filtering scheme in this case introduces logical transformations on the bit pattern by performing an XOR operation. Two consecutives '1s' or '0s' at the input of the DI generate a logical '0' at the output. On the other hand, bit patterns of '0 1' and '1 0' lead to logical '−1' and '1', respectively. The logic operation performed by the DI is illustrated in Fig. 5.9(d). This code transformation can be restored by an electronic differential encoder, see Fig. 5.9(e). To prove this we note that the two devices will perform complementary modulo two operations [202]

$$b_k = b_{k-1} + a_k \tag{5.2}$$

$$\tilde{a}_k = \tilde{b}_k + \tilde{b}_{k-1} \tag{5.3}$$

where a_k, b_k, $\tilde{a}_k$, $\tilde{b}_k$ $\in$ $\{0,1\}$, i.e., $0 + 1 = 1 + 0 = 1$ and $0 + 0 = 1 + 1 = 0$. Substituting (5.2) to (5.3), it can be easily shown that the output of a differential encoder followed by a DI returns identity if b_k is identical to $\tilde{b}_k$

$$\tilde{a}_k = \tilde{b}_k + \tilde{b}_{k-1} = b_{k-1} + a_k + \tilde{b}_{k-1} = a_k \ . \tag{5.4}$$

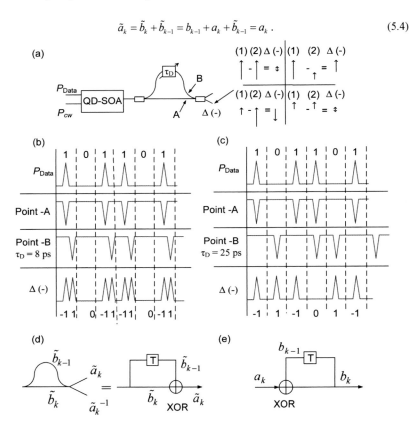

Fig. 5.9: Operating principle of the delay interferometer scheme for wavelength conversion with QD SOA in the high injection regime. (a) Set up and amplitude dependency at the output of the DI of the fields in path (A) and (B). (b) For $\tau_D = 8$ ps the scheme generates a stream with duplicated pulses within each bit-slot, which is inappropriate for transmission. (c) Pulse duplication is avoided when $\tau_D = 25$ ps, i.e., the duration of a single bit at 40 Gbit/s. In that case the scheme generates logical transformation on the output signal with an AMI format. (d) Logic operation performed by of the optical DI. (e) Logic operation performed by the differential encoder, [J7].

In a transmission link, an equal number of differential encoders and DI are required to cancel each other's operation. A digital circuit located either at the transmitter or receiver can be implemented to perform the required number of encoding operations. In a reconfigurable network, that number is not fixed but depends on the physical connection each time is established. There can be different ways to transfer this critical information to the corresponding encoder. For example, in an optical circuit switched network this number can be given from the control plane as a part of the path establishment process. In an optical packet switched network the information can be extracted through appropriate pattern recognition on a predetermined bit phrase of the packet's header [203]. It should be mentioned that the converted output signal of the DI is an alternate mark inversion (AMI) format with phase changes of π for each subsequent '1', irrespective of the number of '0' in-between. Advantages of AMI formats over other binary formats are lack of discrete spectral components, which allows larger launch powers without degradations from Brillouin backscattering, and a favorable suppression of intra-channel FWM as well as reduced XPM, all of which usually results in longer transmission distances [204].

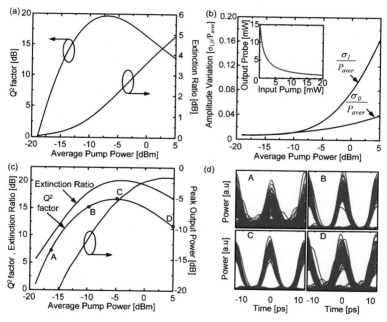

Fig. 5.10: Performance evaluation of the wavelength converter in the regime of high carrier injection. (a) Q^2 factor and extinction ratio of the signal just after the QD SOA as a function of the average pump power that enters the device. Despite the high Q^2 factor the signal suffers from severe extinction ratio degradation. (b) Inset: XGM-based power transfer function of the QD SOA. Amplitude variations (normalized) of the converted signal as a function of the pump power. (c) Q^2 factor and extinction ratio of the signal at the total output of the proposed scheme. Significant extinction ratio improvement is noticed. (d) Corresponding eye diagrams at operating points (A), (B), (C), (D) of subfigure (c), [J7].

Through simulation analysis a detailed evaluation of the QD SOA and the 1-bit DI scheme are evaluated, see Fig. 5.10(a)-(d). First the quality of the signal just after the QD device without any subsequent filter is discussed. Fig. 5.10(a) illustrates the corresponding Q^2 factor and the extinction ratio as a function of the average in-fiber input pump power on the cw wavelength. For low power levels the data signal cannot saturate the gain, a fact which leads to small extinction ratios (< 1 dB) and Q^2 factors (< 10 dB). By increasing the signal (pump) power the depletion of the QD states becomes stronger, which enhances the extinction ratio and consequently improves the Q^2 factor. For a pump level of –7 dBm the Q^2 factor becomes maximum (~ 19 dB). However, for even larger power levels the Q^2 factor starts degrading again although the extinction ratio does not decrease. The explanation comes from the noise transfer function properties of the QD SOA, which do not differ from a bulk SOA [205], [206]. The power transfer function of the XGM process, inset in Fig. 5.10(b), suggests that the amplitude fluctuations superimposed to incoming '1s' are suppressed, while those on the incoming '0s' are enhanced. At the input of the device we assumed a data signal with a constant Q^2 factor of 21 dB. This implies that the amplitude fluctuations for both '1's' and '0's' vary linearly with the average input power level. Fig. 5.10(b) shows the corresponding fluctuations of the in-fiber output probe signal, in terms of the standard deviations normalized to a constant in-fiber power level of –8 dBm. At low pump powers, the input noise is limited, and the device transfers minor distortions to the converted signal. For higher power levels the input noise is increased, so the SOA produces highly distorted '1's', while it suppresses the fluctuations of the '0's'. The increasing noise distortion on the output '1's' is responsible for the Q^2 factor degradation to be noticed in Fig. 5.10(a).

Although the converted signal after the QD SOA can take high Q^2 factor values, it is hardly of any practical use due to the signal's low extinction ratio (< 2 dB). Adding the 1-bit DI filter will significantly improve this situation. Fig. 5.10(c) depicts the Q^2 factor, the extinction ratio and the peak power of the signal at the output of the proposed scheme as a function of the average pump power. As the 1-bit DI subtracts the field amplitudes of consecutive bits, the extinction ratio of the signal after the QD device will define the peak power at the final output. At low pump powers, the resulting output is close to the noise level. Consequently, a large extinction ratio and Q^2 factor degradation take place. When increasing the pump level, the total output power monotonically increases. In that case, the signal extinction ratio improves significantly, approaching 20 dB. The Q^2 factor increases as well and takes on its maximum value 16.5 dB when the pump power is –6 dBm. Beyond this point the signal quality degrades again due to the QD SOA's aforementioned inefficiency in noise suppression.

It should also be pointed out that the DI degrades the OSNR of the converted signal. This is attributed to the 1-bit differential functioning of the filter, which subtracts between two consecutive bits only the coherent part of the optical signal. Its noisy fluctuations, due to their incoherent nature, are practically superimposed as a power average. This OSNR degradation is responsible for the Q^2 factor penalty at the output. To validate our previous conclusions we illustrate in Fig. 5.10(d) the eye diagrams (A), (B), (C), (D) of the final signal that correspond to pump power levels of –17 dBm, –10 dBm, –5 dBm and 5 dBm, respectively. Eye diagrams (A), (D) and partially (B) suffer from ASE noise, while diagram (C) looks much better as it represents operation at the optimum point (see also Fig. 5.10(c)).

Plasma induced phase effects were not taken into account in the aforementioned analysis. Based on the theory presented in [117], it can be seen that the plasma effect influence was minor for the 2 mm long device case which is examined here. Once proper design rules and development techniques are identified to enhance their contribution in future QD SOA it can be believed that those phenomena can also play a key role for ultra-fast switching.

5.3.2 Experimental Realization of Wavelength Conversion

In this subsection the experimental realization of wavelength conversion using QD SOA is shown. The experiments for the proposed schemes have been performed with the 1.55 µm QD SOA (SOA 5) characterized in *Chapter 3*. Additionally performed high-speed wavelength conversion experiments are done with another QD SOA (SOA 3) at a wavelength of 1.3 µm.

5.3.2.1 WC in Low Carrier Injection Regime with 1.55 µm QD SOA

First, the wavelength conversion with the SOA 5 is presented for the case of low carrier injection. Here, the data signal launched into the QD SOA consists of a 42.7 Gbit/s RZ (33 %) OOK at a wavelength of 1550 nm modulated by a PRBS of length $2^{11}-1$. The cw channel is located at 1545 nm. The average in-fiber input power of the data channel is −5 dBm, and the cw in-fiber input power is 0 dBm. The converted signal just after the SOA enters the two DI, where proper tuning of their phases at the operating point maximizes the signal quality.

The result of this process is illustrated in Fig. 5.11(a), (b). Fig. 5.11(a) shows the spectrum of the modulated probe channel, obtained after the QD SOA (black solid line) and after the optical filtering structure (red solid line), respectively. The transfer function of the filter (blue solid line) is shown as well. The signal after the SOA exhibits more blue than red spectral components, which accordingly favors the implementation of blue shifted filtering.

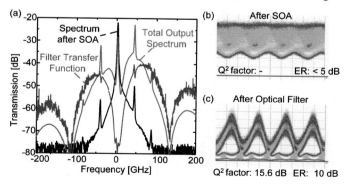

Fig. 5.11: Experimental demonstration of the proposed scheme in the regime of low carrier injection. (a) Spectrums of the modulated probe channel just after the QD SOA (black) and after the optical filter (red) at the optimum operating point. The transfer function of the filter is also shown (blue). (b) Eye diagram after SOA and (c) eye diagram after SOA and filter. It can be seen that the filter significantly improves the quality of the output signal, [J7].

Comparing to Fig. 5.7(c) it can be seen that the experimental results have quite remarkably verified the accuracy of the theoretical predictions for the low carrier injection regime. The eye diagrams of the probe channel behind the QD SOA and the filtering structure are both shown in Fig. 5.11(b) and (c). The inverted signal right after the QD SOA has a signal quality which is indeterminable. However, the signal after the QD SOA followed by the optical filter shows a non-inverted waveform characterized by a good Q^2 factor of 15.6 dB and an extinction ratio of 10 dB. These are the best performance merits and have been achieved for the aforementioned combination of modulating data and cw-probe power levels. Increasing the data power the nonlinear nature of bit-patterning becomes stronger and it cannot be efficiently compensated by a linear optical filtering scheme. On the other hand, at lower data powers the resulted modulation depth is not differentiated from the emitted ASE noise. Therefore, in both cases the Q^2 factor degrades. Finally, the cw-probe power has to be high to enhance the carrier density dynamics in this low carrier injection operating regime.

5.3.2.2 WC in High Carrier Injection Regime with 1.55 μm QD SOA

Demonstration of wavelength conversion in the high carrier injection regime is not possible with today's devices, unless advanced temperature controlling techniques are used [178]. Therefore, to prove the concept of the proposed notch-filtering scheme, the experiment is performed at a low bit rate of 12 Gbit/s RZ OOK to avoid the bit-patterning phenomena, which do not occur for high carrier injection.

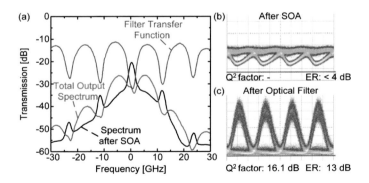

Fig. 5.12: Experimental demonstration of wavelength conversion in the regime of high carrier injection. (a) Spectra of the converted signal at the output of the QD SOA (black curve), and at the output of the 1-bit DI (red curve), respectively. The transfer function of the single DI is also shown (blue curve). Corresponding eye diagrams, (b) after the SOA, and (c) after the single DI. The proposed filtering scheme leads to a significant improvement of the extinction ratio ER, [C28].

The corresponding results are demonstrated in Fig. 5.12(a)-(c). Fig. 5.12(a) depicts the signal spectra after the QD-device (black solid line) and the one after the 1-bit DI (red solid line), respectively, as well as the transfer function of the DI (blue solid line). A pattern-effect free signal with a poor extinction ratio (ER) has been measured just after the QD SOA, see Fig. 5.12(b). However, the 1-bit differential operation that is introduced on the signal by the DI can significantly enhance the ER. As it is shown in Fig. 5.12(c) an improved ER of 13 dB is measured at the output of the subsystem. In addition, the Q^2 factor equals 16.1 dB. The output signal spectrum in Fig. 5.12(a) proves the conversion of the format to an AMI.

5.3.2.3 WC in Intermediate Carrier Injection Regime with 1.3 μm QD SOA

In this section, error-free wavelength conversion of 40 Gbit/s and 80 Gbit/s RZ OOK signals using a QD SOA and subsequent filtering with a DI is demonstrated. A fast gain response in the range of a few ps is found for the QD SOA resulting in open eye diagrams. However, a weak contribution from slow phase effects prevents the use of a 1-bit DI to simply re-invert the XGM signal. Instead, the patterning effects induced by XPM require the use of a DI with a small time offset. It is estimated that the minimum QD filling factor is about 0.7, see Fig. 5.5.

World-record error-free wavelength conversion with a QD SOA combined with a DI was shown for the first time by bit-error ratio measurements at 80 Gbit/s in early 2011 by the author [J4] (the results are presented here). However, it needs to be mentioned that wavelength conversion via XGM in QD SOA was demonstrated by eye diagram quality measurements (no bit-error ratio (BER) measurement) at 40 Gbit/s [207], at 80 Gbit/s for multicast conversion [208], and at 160 Gbit/s [182]. Recently (end of 2011) a wavelength converter based on XGM and XPM with a QD SOA and following filtering has been demonstrated at a bit rate of 320 Gbit/s with good signal qualities also verified by BER measurements [68].

Experimental Setup

The setup to perform wavelength conversion measurements at 40 Gbit/s or 80 Gbit/s RZ OOK data signals consists of a tunable mode-locked laser (TMLL), see Fig. 5.13, emitting a pulse train at 9.953 GHz repetition rate at a wavelength of 1310 nm. The pulse train is multiplexed to 39.812 GHz using a dual stage optical time division multiplexer (OMUX). Quantum-well (QW) SOA boost the pulse train to compensate for the losses induced by the multiplexer stages. A 2^{31}-1 PRBS is imposed on the pulse train via a Mach-Zehnder modulator (MZM). A second OMUX can be used to double the bit rate from 39.8 Gbit/s (~ 40 Gbit/s) to 79.6 Gbit/s (~ 80 Gbit/s). Two praseodymium-doped fiber amplifiers (PDFAs) amplify the data signal to a maximum power of 12.5 dBm at the QD SOA input. The PDFAs ensure pattern-effect free amplification of the modulated data signal. The pulse FWHM is 2.0 ps at the input of the QD SOA. The data signal is combined with a cw probe signal from a wavelength tunable external cavity laser (ECL). The cw probe power launched into the QD SOA is 9.5 dBm.

The wavelength conversion takes place in the QD SOA and in consequence the inverted modulation is transferred by means of XGM and XPM onto the probe. The filter after the QD SOA is used to block the pump signal. The probe signal is explicitly not offset filtered. After amplification in a PDFA, the wavelength converted output of the QD SOA is re-inverted via a tunable DI and analyzed by eye diagram and bit-error ratio measurements.

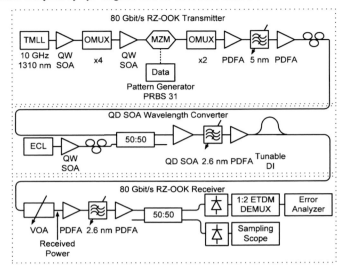

Fig. 5.13: Sketch of the setup for 80 Gbit/s RZ OOK wavelength conversion. TMLL: tunable mode-locked laser, QW SOA: quantum well semiconductor optical amplifier, OMUX: optical time division multiplexer, MZM: Mach-Zehnder modulator, PDFA: praseodymium-doped fiber amplifier, VOA: variable optical attenuator, ECL: external cavity laser, DI: delay interferometer, and ETDM DEMUX: electrical time division demultiplexer, [J4].

The 80 Gbit/s receiver consists of two PDFAs in combination with a 50 GHz photodiode (u^2t XPDV 2320R). In the case of measuring 80 Gbit/s signals the output of the photodiode is demultiplexed in the electrical domain to 40 Gbit/s in front of the error analyzer.

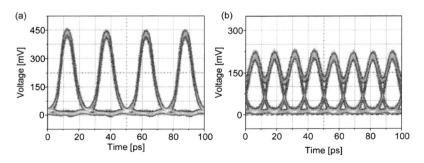

Fig. 5.14: Back-to-back eye diagrams at 1310 nm of the PRBS 2^{31}-1 RZ OOK data signal at (a) 40 Gbit/s and (b) 80 Gbit/s, [J4].

The back-to-back eye diagrams for maximum receiver input power are shown in Fig. 5.14 for (a) 40 Gbit/s and (b) 80 Gbit/s. The limited temporal resolution of the photodiode and the autocorrelation measurements at 80 Gbit/s have proven the pulses to be clearly distinct.

The nonlinear element mediating wavelength conversion is the SOA 3 (QD SOA). At 1310 nm, an unsaturated FtF gain of 7 dB and 5 dB at 250 mA and 500 mA, respectively, has been observed (Fig. 5.15(a)). Gain is predominantly provided for the TE polarization (TE/TM ~ 10 dB). Thus, polarization controllers are required in front of the QD SOA. The QD SOA is operated up to a current of 600 mA, presently limited by heating of the device, which is responsible for the decrease of the unsaturated FtF gain, if the current is increased from 250 mA to 500 mA. Nevertheless, a large current is desirable, since it is the decisive parameter to achieve high speed gain dynamics and thus to minimize patterning.

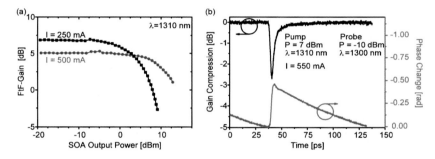

Fig. 5.15: (a) FtF gain of the QD SOA versus in-fiber output power at 250 mA (black) and 500 mA (blue). (b) Pump-probe trace showing gain and phase recovery at 550 mA using 1.6 ps pump pulses at 10 GHz repetition rate, [J4].

Pump-probe traces of the gain and the phase recovery are shown in Fig. 5.15(b). The power of the 10 GHz pump pulse train at 1310 nm is chosen to have roughly the same pulse energy as in the 80 Gbit/s wavelength conversion measurement described in the following. The measurement demonstrates the fast gain dynamics with a 90 %-10 %-recovery time of less than 10 ps. In contrast to the fast gain recovery, the phase does not fully recover within the pulse period of 100 ps. For similar QD SOA a 90 %-10 % phase recovery time as large as 500 ps was observed. A peak-to-peak phase change of 0.45 rad is found.

Experimental Results

Eye diagrams of the converted signal right behind the QD SOA at 1320 nm are presented in Fig. 5.16(a) for 40 Gbit/s and in Fig. 5.17(a) for 80 Gbit/s, respectively, which are recorded without the DI in use. The fast gain dynamics manifests itself in a low broadening of the mark level of the inverted signal and in a reasonable eye opening, however with a low extinction ratio of approximately 2 dB at 40 Gbit/s (1.5 dB at 80 Gbit/s) as expected from the gain compression shown in Fig. 5.15(b). The eye opening demonstrates that the gain dynamics is fast enough for 80 Gbit/s RZ OOK signal processing.

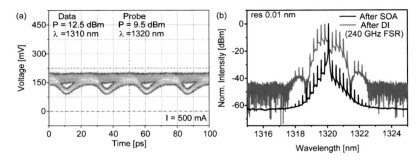

Fig. 5.16: (a) Eye diagram of the converted 40 Gbit/s signal at 1320 nm right behind the QD SOA without the DI in use. (b) Optical spectrum after the QD SOA (black) and after the DI with 240 GHz FSR (red), [J4].

The optical output spectra of the QD SOA and the DI at 40 Gbit/s (Fig. 5.16(b)) and 80 Gbit/s (Fig. 5.17(b)) provide indication for significant phase effects, since the intensity of sidebands with respect to the cw carrier is asymmetric. At 80 Gbit/s, the first left sideband is 4.3 dB larger than the first right sideband. The modulation sidebands are 23 dB below the cw carrier, which explains the poor extinction ratio of the converted signal right at the output of the QD SOA seen in Fig. 5.16(a) and Fig. 5.17(a). These observations imply in combination with the phase recovery found in the pump-probe measurement (Fig. 5.15(b)) that the QD SOA induces in addition to XGM also a significant amount of XPM. As shown in Fig. 5.15(b), the phase recovery takes place on a time scale exceeding 100 ps which is thus much slower than the gain dynamics. The slow refilling of the reservoir in the higher energy states dominates the phase dynamics rather than the QD ground state itself, as shown by simulations in [209].

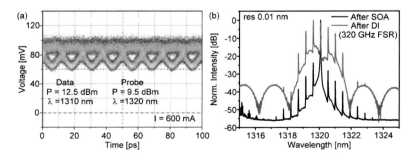

Fig. 5.17: (a) Eye diagram of the converted 80 Gbit/s signal at 1320 nm after the QD SOA without the DI. (b) Normalized optical spectra of the QD SOA output (black) and after the DI (red) for conversion towards 1320 nm. The first right sideband is weaker by 4.3 dB than the first left sideband, [J4].

The phase patterning becomes important, if the QD SOA output signal is filtered with the DI, since it processes both, amplitude and phase. Setting the FSR of the DI equal to the bit rate results in a strongly deteriorated and closed eye diagram of the DI output as shown in Fig. 5.18 for the 80 Gbit/s data signal. In *Section 5.3.1.2* only amplitude modulation due to spectral hole burning was taken into account, but although phase effects of present QD SOA are rather weak, they significantly contribute to the filtered output and can obviously not be neglected in this high-speed experiment.

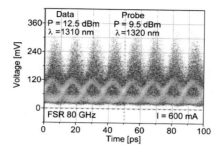

Fig. 5.18: Eye diagram of the converted 80 Gbit/s signal after the DI with a FSR of 80 GHz, [J4].

The principle of signal processing by the DI is sketched in Fig. 5.19. The choice of a suitable FSR enables compensation of the phase patterning. If the FSR of the DI equals the bit rate (Fig. 5.19(a)), successive pulses interfere at the output of the DI.

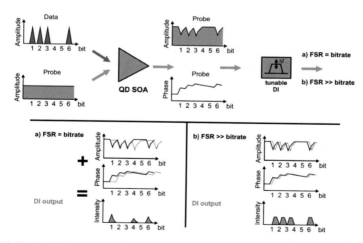

Fig. 5.19: Sketch of the signal processing by the QD SOA and the DI. The QD SOA induces XGM and slow XPM. The non-recovered phase results in patterning effects, if the FSR equals the bit rate. Wavelength conversion with low patterning due to compensation of phase patterning is feasible at large FSR values. The dotted line represents the delayed copy of the solid trace, [J4].

Although the gain is fast enough to recover sufficiently between successive pulses, the phase recovers on a time scale much longer than the bit-period. Since the absolute phase and the phase difference between the pulses strongly depend on the preceding data sequence, the uncompensated phase patterning results in a strongly deteriorated output eye diagram.

In contrast to that, a FSR larger than the bit rate (Fig. 5.19(b)) causes the pulse to interfere with its time-delayed copy. A larger FSR corresponds to a shorter temporal delay, which in turn determines a temporal switching window. The use of a switching window allows mitigation of the phase induced patterning effects, since the phase difference of the two copies of the same bit is less dependent on the preceding bits than the phase difference of neighboring bits. The corresponding optimum delay in the experiments of 4.2 ps (240 GHz FSR) at 40 Gbit/s and 3.1 ps (320 GHz FSR) at 80 Gbit/s represents a tradeoff between the formation of sufficiently short pulses and the output power of the DI. Moreover, this scheme relies on a slow phase recovery, since the difference of the phase of the initial signal and its time delayed copy has to be sufficiently small after the initial switching window (i.e., beyond the delay time) to prevent the generation of a second switching window. Otherwise, trailing pulses are generated which may cause intersymbol interference. Autocorrelation measurements of the 80 Gbit/s DI output signal for a FSR of 320 GHz exhibit a triangular shape indicating square-like pulses in the time domain.

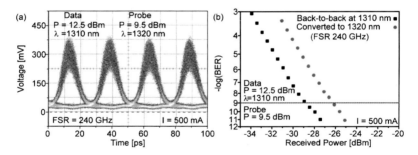

Fig. 5.20: (a) Eye diagram after DI for maximum receiver input power. (b) Bit-error ratio versus received power for 40 Gbit/s wavelength conversion from 1310 nm to 1320 nm, [J4].

In the wavelength conversion experiments shown here, the DI is adjusted for maximum suppression of the cw carrier. FSR values of 240 GHz and 320 GHz are chosen to filter the 40 Gbit/s and 80 Gbit/s data signals, respectively. The spectra of the DI output at 40 Gbit/s in Fig. 5.16(b) (at 80 Gbit/s in Fig. 5.17(b)) prove that the asymmetry of the sidebands is unaffected by filtering, whereas the sideband-to-carrier ratio is tremendously enhanced to 2.3 dB at 40 Gbit/s (1.7 dB at 80 Gbit/s) by means of the DI. The large FSR values result in a significant improvement of the output eye diagrams of the wavelength converter at 40 Gbit/s (Fig. 5.20(a)) and 80 Gbit/s (Fig. 5.21(a)) for conversion from 1310 nm to 1320 nm. Extinction ratios of 10 dB and 9.6 dB are measured at 40 Gbit/s and 80 Gbit/s, respectively.

The tunable band-pass filters were not used as offset filters. BER measurements of the 40 Gbit/s output signal of the wavelength converter (conversion from 1310 nm to 1320 nm) demonstrate error-free performance without an error floor down to 10^{-12} (Fig. 5.20(b)). In comparison to the back-to-back measurement a penalty of 2.5 dB is found at a BER of 10^{-9}. The reduced receiver sensitivity at 1320 nm compared to 1310 nm causes 0.9 dB penalty, the remaining penalty of 1.6 dB could be attributed to not fully compensated phase patterning and to the lower extinction ratio of the signal at the output of the wavelength converter.

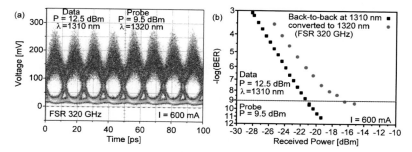

Fig. 5.21: (a) Eye diagram of the converted 80 Gbit/s data signal at 1320 nm after the DI with 320 GHz FSR. The extinction ratio is 9.3 dB. (b) BER versus received power for 80 Gbit/s RZ OOK wavelength conversion from 1310 nm to 1320 nm, [J4].

Wavelength conversion of an 80 Gbit/s RZ OOK signal (Fig. 5.21 (b)) from 1310 nm towards 1320 nm shows error-free (BER < 10^{-9}) operation. A penalty of 4.8 dB is found compared to the back-to-back measurement. However, an error floor appears at a BER of about 10^{-10}. OSNR measurements confirm that this error floor in the BER measurement versus received optical power is caused by a limited OSNR after the wavelength converter. Neglecting the error floor yields an estimated power penalty of about 2.5 dB at 10^{-9}. Similar to the measurements at 40 Gbit/s, this includes 0.9 dB penalty from the reduced receiver sensitivity at the wavelength of the converted signal. The remaining penalty of about 1.6 dB can again be attributed to the limited extinction ratio and the uncompensated phase patterning which results in an amplitude patterning effect due to phase-to-amplitude conversion in the DI.

The measurements presented here demonstrate that fast gain dynamics of present QD SOA enable wavelength conversion at high bit rates with low amplitude patterning. However, a filter, here a DI, is used to re-invert the signal and to improve the extinction ratio. The important finding is, that the DI has to be operated at a large FSR in order to mitigate phase patterning effects resulting from the slow carrier dynamics of the reservoir. Although the phase effects in QD SOA are commonly found to be smaller than in conventional devices, they are still sufficient to allow phase based wavelength conversion. Therefore, for wavelength conversion using present QD SOA the slow phase dynamics cannot be neglected and use of the fast gain dynamics only is not sufficient for high speed wavelength conversion, similar to the conversion using conventional amplifiers [210].

The advantages of the wavelength converter used in this experiment are, that filtering with a DI avoids a format conversion for re-inversion of the converted signal and does not require an additional offset filter. A QD SOA and a single DI could easily be integrated into a single chip wavelength converter.

5.3.3 Design Strategies for Ultra-fast and Nonlinear QD SOA

Operating next-generation QD SOA in the high carrier injection regime without any contributions from phase effects requires a simpler filtering scheme. From a theoretical point of view it is always possible to achieve complete population inversion between wetting layer and QD ground states by increasing the injected current density beyond a specific limit. However, most of the times this theoretical limit is far from what a real device tolerates. Design rules for QD SOA that achieve complete population inversion at much lower current densities are therefore required.

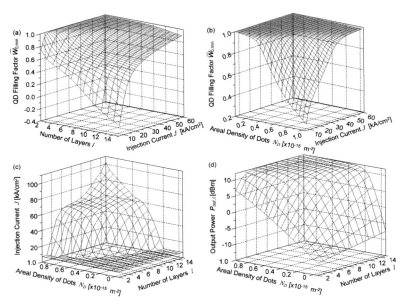

Fig. 5.22: Design rules of the QD SOA for operation in the high carrier injection regime. (a) QD filling factor as a function of the number of layers l and the injection current J. (b) QD filling factor as a function of the areal density of dots N_D and the injection current J. (c) Required current density to achieve complete population inversion as a function of the areal density of dots and the number of layers. Reducing both parameters lower current density values are required. (d) Average power of the converted signal as a function of the areal density of dots and of the number of layers, [J7].

A straightforward approach would be to decrease the number of available QD states. This would require a reduced number of reservoir (WL) carriers to fill the empty states of the QD.

Such a scenario is analyzed in Fig. 5.22(a)-(d). The total number of QD states depends on the number of layers l and on the areal density of dots N_D, see QD model in *Section 2.3.3*. Fig. 5.22(a) illustrates the QD filling factor $\overline{W}_{0,min}$ as a function of the number of QD layers l and of the injected current density J. The areal density N_D of dots is equal to the nominal value of 0.79×10^{-15} m^{-2}, which corresponds to the SOA 5 (see Table 5.1). The average in-fiber input power levels of the cw and data channels were set at 0 dBm and -5 dBm, respectively. The figure shows that for a large number of layers, i.e., $l \geq 10$, high current densities, i.e., $J \geq 60$ kA / cm^2, are required for full population inversion. On the other hand, for $l \leq 4$, operation in the higher injection regime can be achieved for current densities lower than 15 kA / cm^2. This value may be considered as an operating limit for a real device.

Similar conclusions can be drawn from Fig. 5.22(b), which depicts the QD filling factor $\overline{W}_{0,min}$ as function of the injected current density and of the areal density of dots. Here, the number of layers equals $l = 6$. Full population inversion $\overline{W}_{0,min} = 1$ at current densities lower than 15 kA / cm^2 is feasible, when N_D does not exceed 0.4×10^{-15} m^{-2}.

Fig. 5.22(c) combines the results of the two previous figures giving a clearer view of the required current densities for complete population inversion. These are depicted as a function of both the number of layers and of the areal density of dots. The figure shows that the areal dot density influences the required current density threshold stronger than the number of layers. Specifically, by decreasing the dot density to less than one half of the nominal value $N_D \leq 0.4 \times 10^{-15}$ m^{-2} while keeping $l \leq 6$, complete population inversion is achieved for current densities within the operating limits of our device (< 15 kA/cm^2).

The total number of dots represents the amount of active SOA material. Therefore, reducing the areal dot density or the number of layers both decreases the gain of the device. Fig. 5.22(d) depicts the average power of the FtF output probe channel as a function of the layer number and the areal dot density. The same pump-probe power levels as previously have been considered at the input of the device. For the nominal values of N_D and l the average in-fiber output power of the probe is 12 dBm. When the areal dot density is reduced to 0.4×10^{-15} m^{-2} the output power decreases to 7.5 dBm, which corresponds to an effective gain of 12.5 dB. Considering that the 1-bit DI after the QD SOA will normally introduce 10 dB of additional power penalty, 2.5 dB of effective gain still remain. Therefore, although with the proposed design strategy the ultra-fast nonlinear performance of the subsystem is optimized, still a positive conversion efficiency factor can be achieved. This factor is much higher comparing to what many alternative schemes have achieved, i.e., using FWM in bulk SOA [211], and it can still ensure robustness against input OSNR degradations even at higher bit rates (> 100 Gbit/s). Alternatively, the reduction of the effective gain can be compensated by increasing the length of the QD device, without canceling its ultra-fast response.

The proposal for the new design approach of the QD SOA when aiming for nonlinear applications has been also evaluated with the 1-bit DI wavelength conversion scheme from *Section 5.3.1.2*.

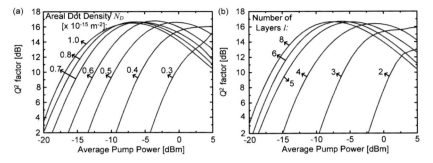

Fig. 5.23: Calculated Q^2 factor curves of the signal at the output of the QD SOA and 1-bit DI scheme as a function of the average pump power in (a) for different areal dot densities, and in (b) for different number of layers. In both cases, the same maximum Q^2 factor value is maintained when the total number of dots is decreased. However, this occurs at higher input power levels as a result of the corresponding increase of the input saturation power, [J7].

Fig. 5.23(a), (b) show the calculated Q^2 factor of the total output signal as a function of the average in-fiber pump power that enters the device. In Fig. 5.23(a) the corresponding curves have been taken for different values of the areal dot density, while $l = 6$ has been chosen. In Fig. 5.23(b) $N_D = 0.8 \times 10^{-15} m^{-2}$ is kept and the number of layers is varied. In both cases, the reduction of the total number of dots has minor influence on the maximum Q^2 factor apart from the fact that it occurs at higher input power levels. This is because the corresponding reduction of the unsaturated gain that takes place effectively increases the input saturation of the device. Therefore, larger signal power levels will be required to achieve the optimum balance between the extinction ratio enhancement and the noise suppression capabilities of the QD SOA for maximizing the Q^2 factor. Increasing the length of the device, however, is expected to counterbalance the corresponding reduction of the unsaturated FtF gain, so that the same maximum Q^2 factor is maintained at the nominal in-fiber signal power levels.

5.4 Multi–Wavelength Clock Recovery with QD SOA

In this section the clock recovery performance for single and multi-channel operation at 40 Gbit/s RZ OOK is experimentally demonstrated with PRBS data patterns of lengths from 2^7-1 up to $2^{31}-1$. The scheme is based on a low finesse Fabry-Pérot (FP) filter followed by a QD SOA. A very high quality clock signal is obtained throughout the C-band with very fast rise and fall times and extremely low rms timing jitter characteristics. The multi-wavelength performance of the circuit is explored and two-channel clock extraction at 20 nm spacing is demonstrated.

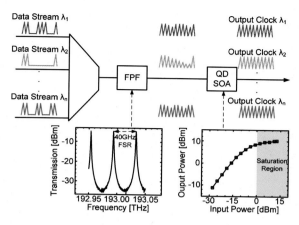

Fig. 5.24: Operating principle of the multi-wavelength clock recovery scheme. The FP filter generates the clock pulses at every wavelength by filling the zeros of the incoming data-stream with preceding ones yielding a clock resembling with strong amplitude variation at the marks stemming from its low decay time. The QD SOA operated as an optical power limiter on a per wavelength basis, performs peak pulse power equalization resulting in high quality output clock extraction.

Operating Principle

The operating principle of the multi-wavelength clock recovery scheme relies on the properties of its building blocks, as illustrated in Fig. 5.24.

The FP filter generates the clock pulses by filling the zeros of the incoming data-stream with preceding ones hence producing a clock-resembling signal with strong amplitude variation at the level of the marks, hereinafter called amplitude modulation (AM). The transfer function of the FP filter illustrated in Fig. 5.24 clearly shows the resonances of the filter spaced at a free-spectral range (FSR) of 40 GHz which exactly matches the bit rate of the incoming data-streams. The low finesse of the filter ensures that the recovered clock signal is extended by only a few bits offering also the potential of efficient clock extraction for packets within a very short guard band.

In turn, the QD SOA performs pulse power equalization of every clock resembling signal at the FP filter output. This is achieved by driving the device into the saturation region where low peak power input pulses experience high gain and high peak power input pulses experience low gain. The result is a recovered clock signal with pulses of equal amplitude and ultra low timing jitter characteristics for every input wavelength.

Multi-wavelength operation of the clock recovery scheme requires the following conditions to be met: (i) the bit rate of all input data signals matches the FSR of the FP filter, (ii) the channel spacing is an integer multiple of the filter FSR so that all wavelength channels coincide with the resonant peaks of the filter, and (iii) the channel spacing is larger than the homogeneous linewidth of the QD to ensure spectral isolation between adjacent channels.

Experimental Setup

The experimental setup is illustrated in Fig. 5.25. Two continuous wave (cw) signals at 1530 nm and 1550 nm are modulated by an electrically generated 40 Gbit/s 33% RZ OOK 2^7-1 PRBS in the transmitter section (Tx). The data-streams are amplified by an erbium doped fiber amplifier (EDFA) and launched into the FP filter (FPF) with 40 GHz FSR and a finesse of 73 (bandwidth of about 500 MHz). Before entering the QD SOA (SOA 5), the two data signals are decorrelated using an optical delay line (DL). The in-fiber power level of every channel before entering the QD SOA is independently controlled by an EDFA followed by a variable optical attenuator (Att.). The polarization state of every channel is independently fixed with a polarization controller (PC) at the QD SOA input with respect to the maximum gain of the SOA device where XGM and XPM effects are maximized. Finally, in the receiver section (Rx) a 3 nm tunable optical band-pass filter is used to receive every recovered clock signal. The quality of the clock pulses in terms of AM, root means square (rms) jitter performance and chirping is measured using a digital communications analyzer (DCA), an electrical spectrum analyzer (ESA) and an optical spectrum analyzer (OSA).

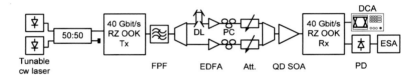

Fig. 5.25: Experimental setup of the multi-wavelength clock recovery circuit. 40 Gbit/s 33% RZ OOK data-streams are launched into a Fabry-Pérot filter (FPF) followed by the quantum-dot SOA (QD SOA). The AM, rms timing jitter and chirping of the recovered clock pulses are measured by a digital communications analyzer (DCA), an electrical spectrum analyzer (ESA) and an optical spectrum analyzer (OSA), respectively, [C14].

Experimental Results

Single channel performance of the clock recovery circuit has been fully characterized and investigated. Fig. 5.26(a)-(c) illustrate the pattern (left), the eye diagram (middle) and the 10 Hz - 100 MHz rf spectrum (right) of the (a) back-to-back (BtB), (b) FPF output and (c) QD SOA output signals, respectively. The AM is measured as the logarithmic ratio of the maximum over the minimum peak pulse power using a 10 % window with respect to the center of the eye diagram and the rms timing jitter is measured by integrating the single sideband (SSB) phase noise around the 40 GHz clock harmonic within 10 Hz - 100 MHz. The used channel average in-fiber input power that deeply saturates the QD SOA is 13 dBm. The clock resembling signal at the FPF output suffers from 5.7 dB AM that is significantly suppressed to only 0.5 dB at the output of the QD SOA. Furthermore, the recovered clock signal has only 120 fs rms timing jitter illustrating its high quality performance characteristics (synthesizer to generate the 40 Gbit/s data signal has a timing jitter of about 140 fs [6]). It is noteworthy that, the above single channel results are obtained at the 1546.5 nm wavelength, with identical performance across the C-band.

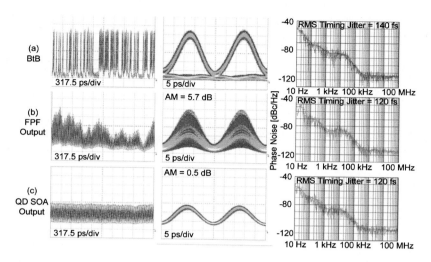

Fig. 5.26: Pattern (left), eye diagram (middle) and 10 Hz - 100 MHz RF spectrum around the 40 GHz clock harmonic (right) of the (a) back-to-back signal, (b) FP filter output, and (c) QD SOA output, for single channel operation at 40 Gbit/s. The RF spectrum for frequencies larger than 10 MHz is dominated by the noise contribution from the electrical spectrum analyzer.

Performance optimization of the multi-wavelength clock recovery circuit highly depends on the average power into the QD SOA. Fig. 5.27(a) illustrates the AM of the recovered clock signals for the single and the two-channel case as a function of the channel input power. A common trend is that the residual peak pulse power fluctuations significantly decrease as the input power approaches the saturation level of the QD SOA. This is more pronounced in the single channel case, where amplitude variation suppression of nearly 5 dB with respect to the FPF output can be achieved. For the two-channel case, the two output clock signals have the same performance around 0 dBm channel input power into the QD SOA for which AM is suppressed by almost 3 dB. Fig. 5.27(b) illustrates the performance of the clock recovery circuit in terms of residual AM for higher PRBS orders in terms the SOA channel input power is always 13 dBm. The clock resembling signal at the FPF output acquires strong peak pulse AM which reaches 7 dB for $2^{31}-1$ PRBS. The latter can be suppressed by up to 6 dB in the single channel case and by up to 2 dB in the two-channel case when considering $2^{31}-1$ PRBS. This highlights the robustness of the scheme to longer sequences of zeros.

Fig. 5.27 shows that the two recovered clock signals at 20 nm wavelength spacing (1530 nm and 1550 nm) with an average SOA channel in-fiber input power of 12 dBm have a residual AM of the recovered clock signals of 2.8 dB and 2.2 dB for the 1530 nm and the 1550 nm wavelengths, respectively. This is indicating that carrier depletion is stronger at the longer wavelength reducing the available gain at the shorter one. Multi-wavelength performance has been investigated for the worst case scenario assuming maximum channel cross-talk by independently adjusting the polarization of every input channel to the maximum output power. Therefore, the per-channel performance is expected to vary from the single channel case. Nevertheless, multi-wavelength clock extraction with low amplitude and rms timing jitter is also achieved. Channel cross-talk arising from carrier interaction among the homogeneous spectral regions of the QD SOA is expected to be alleviated if wavelength spacing increases beyond 20 nm or if the material properties of the QD are properly tailored to enhance spectral isolation among adjacent homogeneous bandwidths [49].

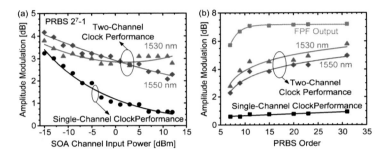

Fig. 5.27: (a) AM of the recovered clock signals as a function of the channel input power for 2^7-1 PRBS and (b) AM of the clock-resembling and the recovered clock signals for various PRBS orders, [C14].

It is noteworthy that the performance of the single and multi-wavelength clock recovery circuit is stable with respect to bit rate and wavelength drifts. The locking range of the clock recovery functionality exceeds 750 ps since the AM is not significantly affected even for a long sequence of zeros at a PRBS of $2^{31}-1$.

5.5 Conclusion and Comparison

In this chapter, desirable parameters for a nonlinear SOA, the wavelength conversion system tests and a multi-wavelength clock-recovery were described. In the following, a summary and a comparison to results of other groups are provided.

Nonlinear QD SOA Wavelength Conversion of High-Speed Intensity-Encoded Signals:
Filtering solutions for all-optical wavelength conversion with QD SOA have been proposed and experimentally verified. In a low carrier injection regime, QD SOA provides a limited amount of cross-phase modulation. The optimum solution is a blue-shifted filtering scheme, implemented by two tunable DI and an OBPF. The filter provides effective mitigation of bit-pattern effects. In a high carrier injection regime, bit-pattern effects and phase effects do not exist. However, a wavelength converted signal suffers from a bad extinction ratio. A notch

filter that will suppress the signal's carrier frequency is considered as the remedy to this deficiency.

The proposed schemes are successfully experimentally demonstrated with 10 Gbit/s, 40 Gbit/s and 80 Gbit/s OOK data signals. World record 80 Gbit/s OOK all-optical wavelength conversion from 1310 nm to 1320 nm using a QD SOA combined with a delay interferometer as subsequent filter has been demonstrated for the first time in the year 2011 with a measured BER $< 10^{-9}$.

In Table 5.2 the best wavelength conversion results based on QD SOA are summarized. First of all, it should be mentioned that this summary focuses on QD SOA and on the OOK modulation format only. The Table 5.2 shows the years in which the wavelength conversion was demonstrated, the bit rate, the used effect (XGM, XPM or FWM), the filtering scheme (DI or OBPF) and the evaluation of the signal quality.

The first 40 Gbit/s OOK (row No. 2) wavelength converter with QD SOA based on XGM and XPM with a measureable signal quality Q^2 of 15.5 dB has been built within this thesis in the year 2008. Additionally, in the year 2011, the author has built the fastest wavelength converter based on XGM and XPM with QD SOA including BER measurements (row No. 8).

Recently, a new world-record for wavelength conversion with QD SOA has been achieved at 320 Gbit/s based on both XGM and XPM (row No. 8).

OOK:

No.	Year 20xx	Bit rate [Gbit/s]	Effect	Filter	BER, Q^2	Ref.
1	02	10	XGM	None	no	[50]
2	03	40	XGM	None	no	[207]
3	08	40	XGM + XPM	DI	15.5	[C33]
4	08	40	XGM + XPM	OBPF	10^{-9}	[181]
5	09	160	XGM + XPM	OBPF	19 dB	[182]
6	10	40	XGM + XPM	2 DI + OBPF	15.6	[J7]
7	11	40	FWM	Dual Pump	10^{-9}	[212]
8	11	80	XGM + XPM	DI	10^{-9}	[J4]
9	11	160	XGM + XGPM	DI	10^{-9}	[68]
10	11	320	XGM + XPM	DI	10^{-9}	[68]

Table 5.2: Comparison and evolution of wavelength conversion speed based on QD SOA for OOK modulation.

Nonlinear Multi-Wavelength QD SOA Clock Recovery: An all-optical multi-wavelength clock recovery based on a Fabry-Pérot filter followed by a QD SOA at a bit rate of 40 Gbit/s has been demonstrated. High quality clock extraction is achieved for single wavelength throughout the C-band. The multi-wavelength performance of the clock recovery can be further improved by engineering the QD SOA.

For the first time, a clock recovery unit based on a FP filter followed by a QD SOA has been implemented in this thesis. However, it should be mentioned that the obtained measurement results are comparable to results for a FP filter followed by a conventional SOA.

Generally, it can be concluded:

- Two operating regimes are present in nonlinear QD SOA: (a) a low carrier injection regime (bit-pattern effect and strong phase contributions occurring) and (b) a high carrier injection regime (no bit-pattern effects and no phase contribution occurring).
- The regimes require different filtering techniques to enable wavelength conversion: (a) a blue-shifted filtering technique based on cross-phase modulation, (b) a notch filtering technique based on cross-gain modulation.
- QD SOA can be used for all-optical nonlinear signal processing at high bit rates of 40 Gbit/s: a wavelength converter and a clock recovery unit have been realized in the regime (a). In this regime, QD SOA behave similar to conventional SOA.
- QD SOA can be used for all-optical nonlinear signal processing at highest bit rates of 80 Gbit/s: a wavelength converter in an intermediate regime between (a) and (b) has been realized. In this regime, bit-pattern effects are absent, but still some phase contributions are present. Thus, a filtering scheme according to the regime (a) is efficiently used to generate phase-to-amplitude conversion.
- Once optimized QD SOA, low number of QD-layers and low areal QD density, are fabricated, they should be able to be efficiently operated in regime (b), and at highest bit rates exceeding 100 Gbit/s. A model experiment and simulation results presented the wavelength conversion performance of the notch-filtering scheme.

6 Metro–Access Switch

Optical cross-connects and optical switches are key devices to enable future all-optical high-speed networks. Especially, aggregation and switching of optical signals are important features currently unavailable in the optical domain.

In this chapter, a regenerative all-optical grooming switch for interconnecting 130 Gbit/s on-off keying (OOK) metro-core ring and 43 Gbit/s OOK metro-access ring networks with switching functionality in time, space and wavelength domain is demonstrated. Key functionalities of the switch are traffic aggregation with time-slot interchanging functionality, optical time-division-multiplexing (OTDM) to wavelength division multiplexing (WDM), so-called OTDM-to-WDM demultiplexing and multi-wavelength 2R regeneration. A micro-electro-mechanical system (MEMS) switch in combination with all-optical wavelength conversion guarantees non-blocking space and wavelength switching for any tributary. Laboratory and field demonstration show the excellent performance of the new concept with error-free signal transmission and Q^2 factors above 20 dB. This work has been performed in the framework of the FP6 European project "TRIUMPH-Transparent Ring Interconnection Using Multi-wavelength PHotonic switches".

Section 6.1 presents the fundamentals. The metro gap and the need for optical grooming switches are presented. *Section 6.2* introduces the optical grooming switch concept. In *Section 6.3*, the challenges using an SOA-based regenerator are discussed. *Section 6.4* shows the switch implementation and the experimental results, and *Section 6.5* concludes the chapter.

The content of the *Sections 6.1, 6.2 and 6.4* has been published by the author in [J5]. Parts of the content of the *Section 6.3* have been published by the author in [C35]. Some changes have been done to adjust the notations of variables, the figure positions and the wording. Additional publications from the author which are used for this chapter: [J8], [J12], [C18], [C22], [C26], [C32], [C36]. Parts of this chapter have already been presented in the work [6].

6.1 Fundamentals

Optical communication networks have undergone a great evolution during the last years due to the enormous growth of IP traffic (see *Chapter 1*). To cope with the bandwidth demand of the users very high capacity long-haul links have been deployed worldwide [169]. Long-haul networks are optimized for optical transmission, and switching of high capacity traffic volume, thanks to innovations especially in WDM technology [170]. Simultaneously, new access technologies are pushing the fiber to the end-user, supporting new large bandwidth applications such as video-on-demand and online-gaming. Access data rates have increased from kbit/s to Mbit/s, and new emerging technologies promise even higher data rates up to Gbit/s per user. Metropolitan area networks (MAN) will need significant improvement in both capacity and functionality in order to cope with the foreseen bandwidth demand [213]. The technology leaps in the backbone and access parts of the network have so far not been matched with progress in the metro part. This is known as the metro gap [214]. The challenge for next generation metro networks is to flexibly aggregate, transmit and switch the high volume traffic (continuous and burst) between the backbone and access networks in a highly cost efficient way [215]. Most metro networks today are of a traditional architecture and consist of synchronous digital hierarchy (SDH)/synchronous optical network (SONET) interconnected rings [216]. SDH/SONET formats were developed when voice was the dominant end-user application. Therefore this format is circuit-switching oriented and most efficient for multiplexing a large number of low rate circuits. The metro network structure comprises SDH/SONET rings that can be subdivided in metro-access rings and metro-core rings. Metro-access rings are also known as edge rings that collect and aggregate the data from the customer sites. Metro-core rings do further data aggregation and then feed them to the long-haul network.

Today, optical wavelength (circuit) routers are able to transparently (non-blocking) switch traffic within the same network. However, switching of data between networks (metro-core and metro-access rings) are performed using costly optical-electrical-optical (OEO) conversion [217], [218]. These electrically switched digital cross-connects (DXC) are able to perform time-slot interchange (TSI) and spatial switching. It should be noted that the use of DXCs at these points of the network aids bandwidth management by providing excellent traffic grooming capabilities. Signal regeneration is taking place at every DXC node.

However, the OEO conversion makes DXCs expensive and complex with large footprints and power consumption [217], [218]. Furthermore, this electronic technology has proved to be very restrictive exhibiting cumbersome provisioning procedures. For example, a bandwidth upgrade for a ring means that all DXCs interfaces have to be upgraded, which is a costly, time consuming and traffic disruptive procedure.

Future transparent optical switches need to provide the functionality already available in the electronic domain such as time, and spatial switching as well as traffic grooming. Optical switches have already shown the potential to overcome the issues of the electrical-based DXCs, especially in terms of bandwidth limitation. Optical processing is also considered to be highly energy-efficient and may lead to switches requiring less overall footprint compared to their electronic counterparts. With optical grooming, transparent interconnection of networks in terms of protocol, format and bit rate can be offered at much higher capacities compared to DXCs.

While current optical nodes enable switching they cannot provide the necessary transparent mechanism for grooming. This actually means that there is a need for optical grooming switches [219], [220] that are also able to aggregate traffic from one network at lower speed traffic to another network with higher speed traffic.

In the following, the concept of such an optical grooming switch is introduced and demonstrated experimentally.

6.2 Switch Concept

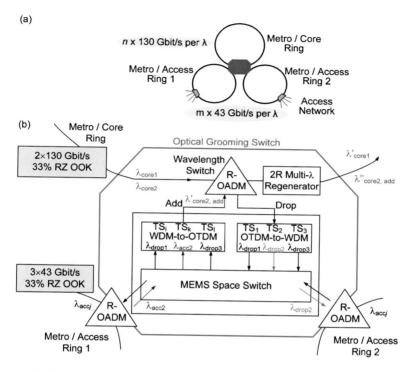

Fig. 6.1: Network scenario and grooming switch. (a) Two metro-access rings are interconnected to a metro-core ring via the grooming switch. Each ring carries a multiple of WDM channels, either at 43 Gbit/s OOK or 130 Gbit/s OOK per wavelength. (b) Grooming switch block diagram, [J5].

The all-optical grooming switch (red octagon in Fig. 6.1(a)) is designed to provide connectivity of a 130 Gbit/s metro-core ring and two 43 Gbit/s metro-access rings. Traffic of three 43 Gbit/s WDM channels is groomed to form a 130 Gbit/s signal in one new WDM channel. In principle, this switch proposal is upgradable to many ports for interconnecting a large number of rings.

Fig. 6.1(b) presents the block diagram of the grooming switch. The metro-core ring carries two WDM channels with 130 Gbit/s OOK data signals (λ_{core1}, λ_{core2}). Each of the 130 Gbit/s signals consists of three OTDM time-slots TS_1, TS_2 and TS_3. One of the 130 Gbit/s metro-core ring data signals is passed through the ROADM and the multi-wavelength 2R regenerator to the output (λ'_{core1}). The other one is first dropped to the OTDM-to-WDM converter, which demultiplexes the OTDM signal with TS_1, TS_2 and TS_3 into three 43 Gbit/s λ-tributaries ($TS_1 \rightarrow \lambda_{drop1}$, $TS_2 \rightarrow \lambda_{drop2}$ and $TS_3 \rightarrow \lambda_{drop3}$). A specific switching scenario for example is that the λ-tributary on wavelength λ_{drop2} is then dropped to access ring 2 and replaced by a new λ-tributary (λ_{acc2}) from access ring 1. λ-tributaries 1 (λ_{drop1}), 3 (λ_{drop3}) and the new λ-tributary (λ_{acc2}) are then directed to the WDM-to-OTDM converter by means of a MEMS. The WDM-to-OTDM unit comprises three asynchronous digital optical regenerator (ADORE) units. Each unit converts one λ-tributary to an OTDM time-slot (i.e., λ_{drop1} → TS$_1$, λ_{acc2} → TS$_2$, and λ_{drop3} → TS$_3$). The OTDM time-slots are interleaved to form the OTDM channel on the new wavelength $\lambda'_{core2,add}$. This signal is finally launched to the metro-core ring through the multi-wavelength 2R regenerator.

The switch offers the following key functionalities:

Traffic grooming [221] is understood here as the aggregation of low-bit rate signals at one wavelength to a high-bit rate signal at a different wavelength, and switching of this signal afterwards.

In this example, the node aggregates 3×43 Gbit/s λ-tributaries to a 130 Gbit/s OTDM signal by utilizing WDM-to-OTDM conversion [222]-[225]. The OTDM signal is then switched to the metro-core ring through a reconfigurable optical add/drop multiplexer (ROADM).

In addition, an OTDM-to-WDM converter [226] enables demultiplexing of a 130 Gbit/s high-bit rate OTDM signal to 3 low-bit rate 43 Gbit/s λ-tributaries.

Time-slot interchanging is the re-allocation of OTDM time-slots of metro-core ring OTDM signals per wavelength. The time-slots of the OTDM signal dropped from the metro-core ring can be interchanged from one time-slot to any other time slot. This is achieved by utilizing OTDM-to-WDM conversion and reconfiguring the MEMS space switch to provide connections to alternate input ports of the WDM-to-OTDM converter.

Optical multi-wavelength 2R regeneration [80], [167] is the simultaneous re-amplification and re-shaping of the various 130 Gbit/s OTDM signals that leave the node via the metro-core ring. In this example, two OTDM signals are considered.

Wavelength selective optical switching is adding / dropping of an OTDM channel per wavelength and the switching of this channel to a specific path of the node. It is implemented with a reconfigurable optical add / drop multiplexer (ROADM) and a MEMS space switch.

6.3 SOA or Fiber–based 2R Regeneration

Optical multi-wavelength 2R regeneration should take place within the all-optical grooming switch. The regenerative subsystem needs to fulfill four requirements.

First, it has to perform re-amplification and re-shaping (2R regeneration) of the incoming data signals. The 2R regenerator should be able to improve the signal quality in terms of extinction ratio (ER) as well as in terms of optical-signal-to-noise ratio (OSNR). Second, the subsystem has to provide wavelength conversion functionality to increase the flexibility of the switch in avoiding wavelength blocking. Third, the subsystem has to simultaneously regenerate data signals on different optical carrier frequencies (wavelengths). Fourth, the subsystem has to cope with high-speed (130 Gbit/s) OOK data signals.

Two technologies have been tested in this work for building a multi-wavelength all-optical 2R regenerative wavelength converter for 130 Gbit/s OOK data signals. A SOA-based regenerative wavelength conversion scheme is compared with a fiber-based scheme. In the end, the scheme with the better performance will be chosen for the all-optical grooming switch implementation.

First, the regenerative capability of a QD SOA wavelength converter is estimated. The wavelength conversion scheme (*Section 5.3.1.2*) consists of a QD SOA with a subsequent delay interferometer (DI) filter. The regenerative capability of this scheme is investigated under the assumption that the QD SOA introduces no bit-pattern effects due to its operation in the high carrier injection regime, see *Section 2.3.2*. A critical element of the proposed wavelength conversion scheme is the DI. Its differential operation can significantly improve the poor ER, which is observed on the wavelength converted signal after the SOA.

The performance enhancement does not necessarily require an SOA with a large alpha factor. Regeneration can also be achieved when the input signal to the DI has obtained a poor phase modulation. In Fig. 6.2(a), an optical pulse-stream is considered at the input of the DI, which has been simultaneously modulated in phase and amplitude (by a virtual SOA). The strength of the optical field amplitude at a logical "1" is A_H, while A_L designates the corresponding field amplitude strength at a logical "0". The efficiency of the proposed wavelength conversion scheme is investigated by calculating the amplitude strength of the optical signal at the destructive Δ-port (normalized to the input amplitude strength) as a function of the input ER and for different phase modulation depths $\Delta\phi$, see Fig. 6.2(b).

For high ER, the "0"-bit wave has a very small amplitude field strength A_L and introduces minor interference to the "1"-bit wave. Thus, the latter will be reduced in the amplitude by a factor of 2 as it propagates through the two cascaded couplers of the DI. At very low ER, the interfering waves have close amplitude levels. If there is no phase difference between them the resulting optical field amplitude at the destructive Δ-port of the DI will be diminished. However, it is important to notice that a small amount of phase modulation ($0.2\,\pi...0.4\,\pi$) is sufficient to keep the amplitude penalty (due to phase-to-amplitude conversion) at values that can be easily compensated by an optical amplifier.

The Q^2 factor performance of the proposed scheme has been examined by simulations. Fig. 6.2(c) illustrates the Q^2 factor improvement calculated at the Δ-port of the DI for different input ER and phase modulation depths. The input OSNR has been degraded by additive Gaussian noise. A constant OSNR (0.1 nm resolution bandwidth) of 20 dB is considered at the input of the DI. On the contrary to the deterministic signal, the power of the uncorrelated ASE noise splits equally at the two ports of the interferometer. Therefore, any improvement on the signal-to-noise ratio will be primarily defined by the amplitude and phase modulation properties of the input pulse-stream. For high ER, the signal power will be degraded by a factor of 4 (two times more than the ASE noise power) resulting in a 3 dB reduction of the corresponding Q^2 factor. As shown in Fig. 6.2(b), phase modulation is not expected to affect that result. On the contrary, due to the strong phase-to-amplitude conversion properties of the DI, the proposed scheme demonstrates a significant capability to improve the quality of the signal waveform that suffers from a bad ER. It can be seen that a large alpha factor inducing a large phase modulation depths is preferable to improve the signal quality. Thus, the use of a high alpha-factor SOA, i. e bulk SOA, is preferable compared to a low alpha-factor SOA, i.e., QD SOA.

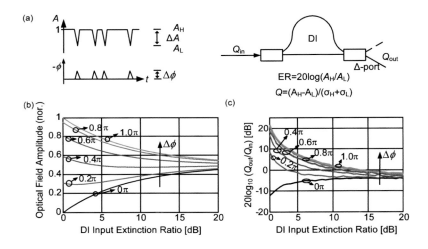

Fig. 6.2: Regenerative capability of the wavelength conversion scheme for high carrier injection (QD SOA operated in the high carrier injection regime and followed by a single delay interferometer (DI) filter). The extinction ratio (ER) is defined after a virtual SOA as ER = $20\log_{10}(A_H / A_L)$, where A_H is the amplitude level of the logical "1" and A_L is the amplitude level of the logical "0". The input OSNR is always constant (resolution bandwidth 0.1 nm) 20 dB. (a) At the input of the DI, an optical pulse-stream is simulated which has been simultaneously modulated in phase $\Delta\phi$ and amplitude ΔA. (b) Normalized amplitude value and (c) Q^2 factor improvement at (destructive) Δ-port of DI as a function of the input ER and for different phase modulation depths $\Delta\phi$, [C28].

Second, the multi-wavelength capability of QD SOA and bulk SOA has been studied. It is found that neither state-of-the-art bulk SOA nor state-of-the-art QD SOA are capable to process nonlinearly multiple wavelengths simultaneously. This is due to strong cross-talk between the data channels. The 3 dB homogeneous linewidth, see Fig. 2.23, in current QD SOA is found to be around 20 nm at room temperature [115] which is about half of the inhomogeneous linewidth. In [59] it has been shown by numerical investigations that a homogeneous linewidth of around 6 nm is required to enable multi-wavelength regenerative wavelength conversion at 160 Gbit/s using a channel spacing of > 10 nm. Even if the homogeneous linewidth is assumed to be 18.6 nm and a channel spacing of 20 nm is used, the SOA introduces strong signal distortions.

Obviously, a SOA-based subsystem is capable for high-speed regenerative wavelength conversion, but it cannot cope with several wavelength channels simultaneously. A possible solution to this issue would be the use of several SOA-based wavelength converters each operating with a single channel.

The lack of a multi-wavelength 2R regenerative subsystem based on SOA requires the use of the fiber-based technology. In principle, a fiber-based solution is transparent to bit rates, formats and their regenerative capability has already been shown [227]. It will be presented in the following section that by a proper choice of the dispersion management a fiber-based regenerator is multi-wavelength capable. The all-optical switch implementation for the grooming switch is thus based on a fiber-based multi-wavelength 2R regenerator operating at 130 Gbit/s on OOK data signals.

6.4 Switch Implementation and Results

In this section, the particular node implementation scenario and the measurement results are presented. The implementation of the subsystems, their performance and the interplay between the different subsystems is addressed.

Following Fig. 6.1(b) and Fig. 6.3(a)-(c), the implementation of the subsystems is as follows:

Reconfigurable Optical Add / Drop Multiplexer (ROADM)

The reconfigurable optical add/drop multiplexer enables wavelength channels to be added (dropped) to (from) the metro-core ring respectively to (from) the metro-access rings. The traffic that is transmitted through the node in the metro-core ring is not affected. In our experiment tunable thin-film filters (TFF) are used.

Wavelength Selective Switch

A ROADM in combination with a space switch (MEMS) is a wavelength selective optical switch (WSS). The ROADM in the metro-core ring selects one of the 130 Gbit/s OTDM signals and drops this channel to the OTDM-to-WDM converter. Following this, the MEMS space switch can redirect each λ-tributary of this signal either to the access ring 1, access ring 2 or back to the metro-core ring. Therefore, in our approach a 1×3 WSS is used.

Wavelength tunability of the ROADM and space switching of the data to the desired output port with the MEMS enable reconfigurable bandwidth allocation. This is required to adapt the network for changing traffic demands of the end-users.

OTDM-to-WDM Converter

An OTDM channel from the metro-core ring may be dropped via the WSS to the OTDM-to-WDM unit.

The OTDM-to-WDM conversion (Fig. 6.3(b)) implies a three stage process. First, the dropped 130 Gbit/s OTDM signal is replicated onto three different wavelengths by means of a multi-wavelength converter. The wavelength conversion is achieved by spectral broadening of the input signal due to SPM within a highly-nonlinear fiber (HNLF) and subsequent filtering at the desired wavelengths (Mamyshev concept [227]). Second, the three replicas are time aligned using an array of optical delay lines so that the respective time slots coincide. Finally, a time gating using an electro-absorption modulator (EAM) extracts every third pulse inside its corresponding WDM channel. The local clock for the EAM gating is provided by a clock recovery unit (CRU). The output 43 Gbit/s λ-tributaries are launched into the MEMS for space switching.

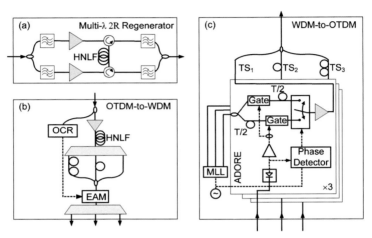

Fig. 6.3: Key building blocks of the grooming switch are the wavelength selective switch (reconfigurable optical add/drop multiplexer (ROADM)), the multi-wavelength 2R regenerator (a), the OTDM-to-WDM converter (b), the MEMS space switch and the WDM-to-OTDM converter (c), [J5].

MEMS Space Switch

Traffic from any of the access rings is switched by means of the MEMS space switch to either metro-access ring or via the add path to the WDM-to-OTDM converter. Each add-port of the MEMS switch relates to a particular time-slot of the OTDM signal. TSI includes the possibility to interchange the time-slots of the OTDM signals per wavelength by dropping and looping-back one wavelength channel through the add path. TSI functionality is thus obtained by reconfiguration of add-ports within the MEMS. In our experiments a 8 × 8 MEMS is used.

WDM-to-OTDM Converter

The WDM-to-OTDM converter (Fig. 6.3(c)) aggregates the three 43 Gbit/s λ-tributaries to a 130 Gbit/s OTDM signal on one wavelength. It consists of three dual-gate ADORE units, each mapping one 43 Gbit/s OOK λ-tributary onto one OTDM time-slot. Each ADORE unit provides regeneration, retiming, pulse width adaptation and wavelength conversion. The OTDM time-slots of the 130 Gbit/s signal are assigned by proper selection of the input λ-tributaries using the MEMS.

The functional principle of the WDM-to-OTDM converter is as follows. Within each ADORE the data signal on the λ-tributary is optical-to-electrical converted within a photodiode (PD) and used to drive two Mach-Zehnder modulators (MZM, gate). The detected signal is also mixed with the local clock to detect the relative phase of the incoming signal. This information is later used to select the correct sampling phase to ensure data integrity between the incoming data on the λ-tributary and the regenerated and retimed signal at the output of the switch. A mode-locked laser (MLL) generates 2.5 ps (full width at half maximum) optical pulses with a repetition rate of 43 GHz which are launched into the three ADORE units. For each ADORE these clock pulses are duplicated and delayed by the time of half a bit-slot. Then the two trains of clock pulses are launched into the MZMs in order to encode the 43 Gbit/s data signal onto the MLL pulse trains. In this way, the incoming data signal is sampled at two points during the bit-slot. Subsequently, one of the data-streams is again delayed by half a bit-slot and the two data signals are directed into a 2 × 1 optical switch. The modulated pulse stream, which is best aligned with the incoming data signal, is selected by a phase comparator circuit. The output signal will therefore be aligned to a fixed output clock phase independently of the incoming data phase. In this way, the random and time-varying bit-slot phases of the input λ-tributaries are translated into a fixed phase. Bit-slips from synchronization onto the common local clock are accommodated within a guard band between bursts, thus maintaining data integrity. A detailed explanation of the experimental implementation can be found in [228]. The tributaries at the output of each ADORE units are subsequently bit-slot interleaved to form the 130 Gbit/s OTDM channel. It should be mentioned that the OEO conversion implemented here, is only used for driving the MZM gates which encodes the bit-stream onto the mode-locked laser pulses. A full photonic WDM-to-OTDM solution seems also to be possible using all-optically switched gates.

Multi-Wavelength 2R Regenerator
To guarantee the quality of the traffic in the metro-core ring, an all-optical multi-wavelength 2R regenerator (Fig. 6.3(a)) operating at 130 Gbit/s is also included. It relies on SPM induced spectral broadening, that takes place in a HNLF, and subsequent filtering at an offset wavelength. This principle is well known for single channel operation and is extended for two wavelengths in this node. To avoid interchannel distortions by cross-phase modulation (XPM) or four-wave mixing (FWM) a bi-directional propagation of the two data signals is used to achieve a rapid "walk-through" of the data pulses within the adjacent channels [80], [167].

Measurement Results
This section will cover the full demonstration of the switch functionalities with multiplexing in wavelength and time. The performance of the solution will be verified by studying a multiple of switching scenarios, showing dynamic bandwidth allocation for time-varying traffic demands. The experimental implementation and results of the switching scenario is shown in Fig. 6.4.

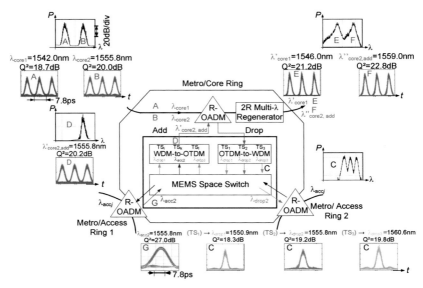

Fig. 6.4: Particular node implementation scenario and corresponding measurement results, where two 130 Gbit/s signals (A & B) are launched into the node. Signal (A) is regenerated and converted to signal (E), signal (B) is split into its 43 Gbit/s λ-tributaries (C) by means of a OTDM-to-WDM converter based on self-phase modulation in a highly nonlinear fiber followed by optical filtering and time gating. λ-tributary λ_{drop2} is dropped to the access ring 2 and a new λ-tributary (G) is added from access ring 1. The two λ-tributaries λ_{drop1} and λ_{drop3} together with the added λ-tributary are aggregated by means of different ADORE units and corresponding time-interleaving. The groomed OTDM signal (see D) is mapped back to the core ring (F) via the 2R multi-wavelength regenerator. The spectra of the signals (A) to (F) are also shown, [J5].

The metro-core ring carriers two 130 Gbit/s signals with λ_{core1} and λ_{core2} with signal qualities of $Q^2 = 18.7$ dB and $Q^2 = 20$ dB, respectively. In a first scenario the 130 Gbit/s metro-core ring signal (A) is passed through the node and the regenerator to the output (E). The eye diagram shows a signal quality improvement to 21.2 dB. The second 130 Gbit/s metro-core ring signal (B) is dropped to the OTDM-to-WDM converter by means of the ROADM. The OTDM-to-WDM converter maps the OTDM time-slots to three λ-tributaries (C) at different wavelengths. The quality of the three λ-tributaries is 19.8 dB, 19.2 dB and 18.3 dB, respectively. λ-tributary 2 with wavelength λ_{drop2} is then dropped to access ring 2 and replaced by a new λ-tributary from access ring 1 with wavelength λ_{acc2} (G). λ-tributary 1 (λ_{drop1}), λ-tributary 3 (λ_{drop3}) and the new λ-tributary are guided into the WDM-to-OTDM unit which generates a 130 Gbit/s OTDM metro-core ring signal on wavelength $\lambda'_{core2,add}$ (D). After regeneration, a high quality 130 Gbit/s signal with a $Q^2 = 22.8$ dB is observed at the output of the switch (F).

Many more switching scenarios are possible. Two other scenarios are considered and the corresponding results are shown in Fig. 6.5. Scenario 2 and 3 show the capability of the switch to perform time-slot interchanging. By reordering the λ-tributaries connections to the

ADORE units through the MEMS switch TSI is achieved. The switching scenarios 2 and 3 are actually identical except for the interchanged time-slot TS_1 and TS_2. The quality of the OTDM channels after the 2R multi-wavelength regenerator is 22.9 dB for scenario 2 and 21.4 dB for scenario 3, respectively. All Q^2 factor measurements are performed with random signal polarization using an all-optical sampling scope.

Switching Scenario	$\lambda^{\cdot}_{core2,\ add}$ Tributaries			$\lambda^{\cdot\cdot}_{core2,\ add}$ Result at F	
	TS_1	TS_2	TS_3	Q^2[dB]	Eye Diagram
1	λ_{-drop1}	λ_{-acc2}	λ_{-drop3}	22.8	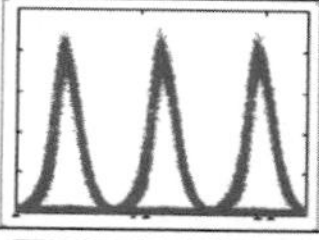
2	λ_{-drop2}	λ_{-acc2}	λ_{-acc3}	22.9	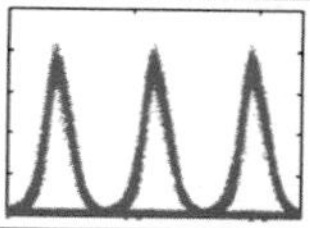
3	λ_{-acc2}	λ_{-drop2}	λ_{-acc3}	21.4	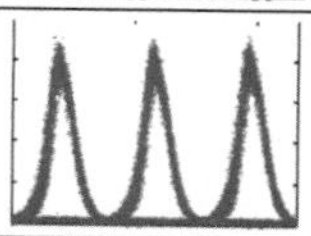

Fig. 6.5: Three switching scenarios with tributaries dropped and looped-back onto the core and access network. Scenario 2 and 3 show time-slot-interchanging. The signal quality of the eyes of the OTDM multiplexed signals is excellent in all situations, [J5].

Field Trial Results

The field experiment aims to demonstrate the key network functions of the switch dealing with impairments introduced in installed fiber links [229]. The field trial (Fig. 6.6(a)) was performed on the Aurora network, an installed dark fiber network within the UK, dedicated to research purposes. For this experiment, two fully dispersion compensated fiber sections were employed. The first section, Colchester-Ipswich-Colchester, was 100% pre-compensated using a slope-matched dispersion compensating module and SMF 28. It had a round trip length of 80 km and represented a metro-access ring with one 43 Gbit/s channel. The second section, Colchester-Chelmsford-Colchester, represented a ring in the metro-core network, had a round trip length of 110 km. It was 80% pre-compensated and 20% post-compensated, and carried two 130 Gbit/s channels.

Several field experiments were implemented. Here only one specific scenario in which the node at Colchester described in Fig. 6.6(b) was separated in two partial nodes is reported. Partial node 1 (Ipswich node) connects the metro-access ring through Ipswich to the metro-core ring through Chelmsford. Partial node 2 (Chelmsford node) drops high-bit rate OTDM signals from the metro-core ring to another access ring. More scenarios can be found in [229].

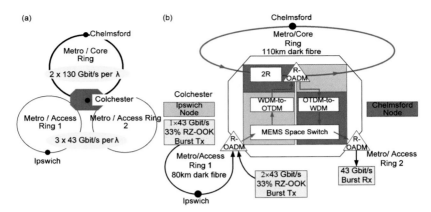

Fig. 6.6: Field trial using an actually installed fiber network. (a) Dispersion compensated dark fiber network between Ipswich, Colchester and Chelmsford in the UK. (b) Network scenario with 1×43 Gbit/s burst traffic transmitted across the 80 km access network to the Ipswich node. Here the traffic is aggregated together with two locally generated 43 Gbit/s burst signals. Time-slot interchanging can be induced depending on the switching scenario. Then the 130 Gbit/s burst traffic has being transported on the metro/core ring over a reach of 110 km to the Chelmsford node where it is 2R multi-wavelength regenerated, demultiplexed and dropped off into the access network, [J5].

In detail, the Ipswich node performs WDM-to-OTDM aggregation of traffic which originates in a 43 Gbit/s edge WDM domain. The Chelmsford node performs 2R multi-wavelength regeneration of two 130 Gbit/s OTDM channels and also OTDM-to-WDM demultiplexing of one of two OTDM channels.

Fig. 6.7: The data frame of the burst switched network consists of a repeating 2^7-1 pseudo random bit sequence (PRBS) of 1 ms duration, and a single modified 2^{19}-1 PRBS of 1 μs duration serving as a guard interval, [J5].

In the experiment one 33% RZ OOK 43 Gbit/s channel is transmitted in access ring 1 through Ipswich to the Ipswich node. Here, another two 43 Gbit/s local channels are launched to the add path. The data pattern (Fig. 6.7) consists of a 2^7-1 pseudo random bit sequence (PRBS) of 1 ms duration and a single modified 2^{19}-1 PRBS 1 μs guard interval with mark-to-space ratio of 52.5 %. This ratio is required to detect the guard interval by observing the change in average power. The data packets and the guard interval were periodically repeated.

WDM-to-OTDM aggregation was performed with the assist of the ADORE unit. The ADORE first detects the guard interval and then performs synchronization of the λ-tributary to the local clock by switching between the two alternate sampling phases. This switching was measured to take place within 440 ns and entirely inside a guard interval. This assured data block integrity during variations of data phase due to runtime differences in the dark fiber.

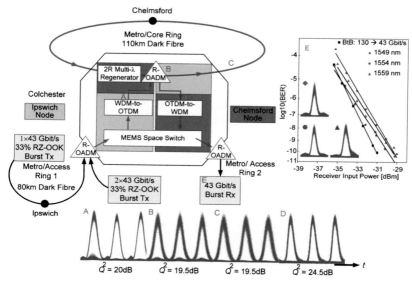

Fig. 6.8: Eye diagrams, signal quality values and bit-error ratio of the field trial measurements at various points in the dark fiber network. Here, the specific scenario in which the Ipswich node performs WDM-to-OTDM conversion of burst traffic which originates in an edge 3×43 Gbit/s WDM domain is reported. The Chelmsford node performs 2R multi-wavelength regeneration of two 130 Gbit/s channels and also OTDM-to-WDM demultiplexing of the OTDM channel. Eye diagram (A) shows the groomed OTDM signal consisting of the output of the ADORE unit combined with two MLL pulses converted onto the same wavelength and interleaved. This OTDM signal is combined with a second 130 Gbit/s OTDM signal (B) which transited the Ipswich node. At the Chelmsford node the two signals are simultaneously regenerated and one channel is dropped (eye (D)) whereas the other channel is passed through the node (C). The OTDM-to-WDM converter generates lower data rate tributaries of 3 × 43 Gbit/s. After space switching the traffic to the access ring, the signal quality of each tributary is measured (E), [J5].

Point (A) in Fig. 6.8 shows the 130 Gbit/s OTDM channel at 1556 nm. It is consists of the ADORE output λ-tributary and two local λ-tributaries. These local signals are pulse width adapted and wavelength converted onto the same MLL as the ADORE output λ-tributary. The generated 130 Gbit/s signal was then combined with another 130 Gbit/s signal (B) which transited the node. Both channels are sent over the metro-core ring through Chelmsford. In the Chelmsford node, they were simultaneously 2R regenerated. One OTDM channel transited the Chelmsford node (C). The second OTDM channel is dropped (D) and OTDM-to-WDM demultiplexed. After space switching the λ-tributary to the access ring, the signal quality is measured (eye diagrams, bit-error ratios (BER) (E)). Excellent eye diagrams and bit-error ratios were measured at all partial nodes of the field experiment. The BER curves in Fig. 6.8

show the results of the 130 Gbit/s-to-43 Gbit/s-EAM-demultiplexed (back-to-back) OTDM channel and the OTDM-WDM converted λ-tributaries. The power penalty is around 2 dB and mainly induced by small leading pulses from the MLL source affecting the CRU performance. The values of the signal qualities and eye diagrams have been measured with an all-optical sampling scope unit.

6.5 Conclusion

A novel all-optical switching node with grooming functionality and multi-wavelength regenerative capability has been successfully demonstrated. The node implementation demonstrated the high quality interoperability of the OTDM-to-WDM, WDM-to-OTDM and 2R regeneration subsystems for continuous traffic. Also, in a field trial using dark fiber links, the tolerance to impairments introduced by fiber transmission together with the switching of data between high bit rate 130 Gbit/s metro-core rings and lower bit rate 43 Gbit/s metro-access rings has been demonstrated with exceptional performance. The node offers switching functionality in time, i.e., including time-slot interchanging, space and wavelength domain. The switch node is expected to boost the progress in the metro networks and to match the technological leaps that have already been carried out at the backbone and the access parts. This approach offers broadband access for every user, and interoperability with existing infrastructures. This provides a smooth migration path from existing to future infrastructures supporting a variety of new services and applications.

7 Summary and Future Work

The global IP traffic is increasing with a growth rate of about 30 % per year. New applications and services such as internet video and video conferencing are the key drivers of the huge bandwidth demand. To satisfy the demand of customers, flexible, reconfigurable, long-reach, non-blocking, high-speed photonic networks are required. The deployment of intelligent all-optical network elements especially in the access and in the metro network can provide this bandwidth. However, an increasing amount of network elements will also increase the total network loss.

Thus, optical amplifiers are required which offer small footprint, low power consumption, are capable of being integrated, and which can amplify all kinds of data signals.

Additionally, there is a need for intelligent signal processing devices to switch and wavelength convert data signals in the optical domain. Especially the transparent switching between networks, which are operated at different bit rates, is a highly desirable functionality. It requires optical grooming switches which aggregate and switch data signals. Such switches include also wavelength converters to avoid wavelength blocking due to wavelength contention.

Linear and nonlinear semiconductor optical amplifiers (SOA) are promising devices for future networks to satisfy the above mentioned demands.

In this thesis, SOA were investigated as amplifiers in their linear gain regime in which the gain is constant. Further, SOA were investigated as signal processing devices in their nonlinear gain regime in which the gain is suppressed. Additionally, an optical grooming switch was studied and implemented.

Especially, novel quantum-dot semiconductor optical amplifiers (QD SOA) which offer unique properties were analyzed. The potential of these devices for all-optical applications such as reach-extender amplifiers in access networks, wavelength converters in switches or clock recovery units were explored. Detailed steady-state and dynamic measurements on QD SOA were performed. The devices were not only tested with traditional on-off-keying data signals, but also with advanced modulation formats, orthogonal frequency-division-multiplexing and radio-over-fiber signals. SOA parameters such as gain, noise figure, saturation input power, SOA dynamics and the alpha factor were optimized with respect to the specific network applications. In high-speed optical communication system measurements, applications of linear and nonlinear SOA, especially QD SOA, were demonstrated.

In this chapter, the key results of this work are summarized and an outlook for future research is provided.

Parameters for Linear Semiconductor Optical Amplifiers

SOA device parameters to design low noise figure and linear SOA were investigated. The SOA parameters were derived in *Section 2.2* and an optimization of these parameters with respect to network applications was discussed in *Section 4.1*.

The quality of signals after an SOA is limited by SOA noise for low input powers, and by signal distortions due to gain suppression for large input powers. Saturation of the gain not only induces amplitude errors but also phase errors due to their coupling by the alpha factor. Saturation also may induce inter-channel cross-talk if several wavelengths are simultaneously amplified by the same SOA.

The lower input power limit is basically determined by amplified spontaneous emission (ASE) noise. The amount of ASE noise added by the SOA is described by its noise figure. Thus, a low input power limit when amplifying data signals with an SOA can be achieved with a low noise figure. The noise figure can be tuned towards the theoretical limit of 3 dB for phase-insensitive amplification using a population inversion factor approaching one. This can be achieved by adapting the current density, i.e., by choosing it as high as possible, with low internal waveguide losses, a good fiber-to-chip coupling, and by choosing the proper dimensionality of the electronic system in the active region, i.e., QD rather than bulk.

The high input power limit is determined by gain saturation induced phase and amplitude changes on the data signal. Thus, a large saturation input power is required to avoid gain saturation. The upper input power limit when amplifying data signals with an SOA can be increased by choosing the gain moderately high, by a large modal cross-section, with a high bias current density, and with a low differential gain. Additionally, if gain saturation cannot be avoided, a low alpha-factor SOA is desirable.

QD SOA show a low amplitude-phase coupling, have an ultra-fast refilling of the QD states, can show large saturation input power levels and introduce only small signal distortions if operated with a high carrier injection rate. In summary, their use as linear amplifiers is strongly suggested.

Input Power Dynamic Range of Linear Quantum-Dot Semiconductor Optical Amplifier

The ratio of a lower and an upper power limit, inside which the reception is approximately "error-free", is expressed in this work by the input power dynamic range (IPDR). Here, the term "error-free" is used for three distinct cases. The first scenario requires a signal quality which corresponds to a bit-error ratio (BER) of 10^{-9}. In the second and third case, the use of an advanced forward error correction (FEC) is assumed, which allows "error-free" operation for a raw BER of 10^{-5} and 10^{-3}, respectively.

The IPDR was first measured with intensity-encoded signals. Largest reported IPDR values about 32 dB with gain > 15 dB at a target BER of 10^{-9} were measured for high-speed on-off keying (OOK) data signals at 40 Gbit/s with optimized QD SOA devices. No additional gain clamping or filtering schemes were applied to extend the IPDR. The complete IPDR study comprises measurements of single and multiple data channels with bit rates from 622 Mbit/s to 40 Gbit/s, see *Section 4.3*.

The IPDR was also investigated for advanced modulation format data signals. Such formats are important for next-generation networks, because data signals in the future will carry more than one bit per symbol. An ideal SOA for advanced optical modulation formats not only needs to properly amplify data signals with various power levels but also should preserve the phase relation between symbols. For the first time, the IPDR of an SOA was systematically studied for high-speed differentially and non-differentially advanced optical modulation format data signals (phase-shift keying and quadrature amplitude modulation) in terms of the alpha factor. It was shown that an SOA with a low alpha factor offers advantages when modulation formats are not too complex. The findings were substantiated by both simulations and experiments performed on SOA with different alpha factors for various advanced modulation formats. In particular, it was shown that the IPDR advantage of QD SOA with a low alpha factor reduces when changing the modulation format from binary phase-shift keying (BPSK, 2QAM; differential: DPSK) to quadrature phase-shift keying (QPSK, 4QAM; differential: DQPSK) and it vanishes completely for 16-ary quadrature amplitude modulation (16QAM). This significant change is due to the smaller probability of large power transitions if the number M of constellation points increases. The smaller probability of transitions with a large power difference in turn leads to reduced phase errors caused by amplitude-phase coupling via the alpha factor. 20 GBd BPSK, QPSK and 16QAM data signals and 28 GBd DQPSK data signals were used in this study. IPDR values of 44 dB for BPSK, 33 dB for QPSK and 13 dB for 16QAM are found, see *Section 4.4*.

Finally, the IPDR was studied for radio-over-fiber (RoF) data signals. It was demonstrated that QD SOA reach extenders provide a large optical IPDR (> 30 dB) as well as a large electrical power dynamic range (EPDR > 30 dB) for 20 Mbit/s QPSK radio-frequency signals on an intermediate frequency of 2 GHz and 5 GHz, see *Section 4.5*.

Applications of Linear Quantum-dot Semiconductor Optical Amplifier

SOA Reach Extender: Linear QD SOA combine ultra-fast dynamics with low distortions from phase changes. Additionally, the moderate gain, the moderate noise figure, the decreased temperature sensitivity of the gain, and the large IPDR make QD SOA a viable solution as in-line amplifiers for next-generation access networks. In-line amplifiers are required to increase the reach as well as the power split ratio to serve a higher number of customers. In such next-generation access networks, a large amplifier IPDR is of particular importance in the upstream path, which carries the data from the customers to the central office. Here, different clients contribute with different power levels due to their different distances to the amplifier.

The application of a QD SOA in-line reach extender for passive optical networks (PON) was experimentally studied in terms of the IPDR. Next-generation long-reach PON requires 1.3 μm and 1.55 μm amplifiers for the upstream and the downstream, respectively. As a result, it was shown that in-line amplifiers based on QD SOA enhance the total loss budget to 45 dB for a large range of input powers. An extended reach PON of 60 km with a split ratio of 1:32 with four downstream wavelength-division-multiplexing (WDM) channels at 2.5 Gbit/s OOK and with two WDM upstream channels at 622 Mbit/s OOK were set up and demonstrated, see

Section 4.6.1. It can be concluded that QD SOA are a promising technology for constructing PON-extender boxes once their polarization dependent gain is reduced in the future.

SOA Loss-Compensating Element: SOA are key devices to overcome loss in next-generation converged metro and access networks. The losses compared to today's networks will increase dramatically due to the insertion loss induced by a growing number of network elements. One of these network elements is a reconfigurable optical add-drop multiplexer (ROADM) which can be used to enable versatile ring network infrastructures. To build metro and access ring networks, which cover the area of a big city, ROADM need to be cascaded to permit a large number of subscribers, e.g., up to 100,000. These ROADM will incorporate loss-compensating SOA, which have to offer highest flexibility in coping with different multiplexing techniques, modulation formats, and optical channel powers. Thus, a study of SOA for metro and access ring networks in terms of multiplexing techniques, modulation formats, and channel power was performed, especially for orthogonal frequency-division-multiplexing (OFDM) data signals. As a result, it was shown that in recirculating fiber-loop experiments with one 1.3 μm QD SOA superior results for OOK signals in terms of distance and IPDR compared to intensity modulated (IM) and direct detected (DD) OFDM with BPSK/QPSK modulated subcarriers are achievable. IM-DD OFDM-BPSK (4.5 Gbit/s including overhead; 8 SOA cascaded with BER $< 10^{-3}$), IM-DD OFDM-QPSK (9 Gbit/s including overhead; 4 SOA cascaded with BER $< 10^{-3}$) data signals as well as 7 Gbit/s and 10 Gbit/s OOK data signals (10 SOA cascaded with BER $< 10^{-3}$) were tested in a 17 km long fiber-loop experiment, see *Section 4.6.3.*

Future Work on Linear Semiconductor Optical Amplifier

In the future, the polarization dependance of QD SOA has to be reduced to achieve a polarization dependent gain around 0.2 dB at a reasonable small-signal gain. A low polarization dependent gain is important for a commercialization of QD SOA for next-generation access networks. Currently, so-called columnar QD SOA with a close stacking of QD layers are used to achieve low polarization dependance of the gain. However, the performance of such devices in terms of gain and phase dynamics as well as the alpha factor is still unknown.

The 60 THz bandwidth offered by the optical fiber will be extensively used by dense wavelength-division-multiplexing (DWDM) in the future. Thus, QD SOA should be grown offering a huge gain bandwidth, preferably in the wavelength range of 800 nm and 1600 nm, in a single device.

Due to the emerging trend in optical communications towards energy efficient and so-called green networks, the required current density to achieve the high-carrier injection regime in QD SOA should be reduced significantly by higher injection efficiency to the QD states.

The amplification of even higher order modulation format signals should be investigated in more detail for, e.g., 8 PSK, 32 QAM and 64 QAM data signals. Interesting is also a study on a WDM performance of QD SOA for BPSK, QPSK, 8PSK, 16QAM and other format data signals especially in terms of the alpha factor. In such a scenario, the influence of the QD SOA alpha factor on the fiber transmission length is also interesting to study. Since a high spectral efficiency is wanted in next-generation networks the influence of four-wave mixing in a WDM approach has to be investigated for advanced modulation formats, too.

Additionally, a concatenation of SOA for advanced modulation format data signals with multiple wavelength channels, with polarization multiplexing and including fiber transmission is of large interest to enable the use of 100 Gbit/s channels over a long distance in the metro and access networks.

Finally, the error-vector magnitude (EVM) as a performance metric has to be further investigated and compared with BER measurement, especially in the nonlinear regime of an SOA.

It should be mentioned that the performance of carefully designed conventional SOA for linear amplification could be made comparable to the QD SOA performance. Therefore, it should be further studied which SOA technology shows the highest potential once the devices are fully optimized for a specific application.

Parameters for Nonlinear Semiconductor Optical Amplifiers

Device parameters to design a highly nonlinear SOA are reported, see *Section 5.2*. The quality of signals after a nonlinear SOA is mostly influenced by the carrier dynamics and the alpha factor. To observe nonlinearities in an SOA, gain suppression is required. In this operating regime, the carrier dynamics should be fast to avoid bit-pattern effects.

The gain of the SOA should be moderate since a minimum of gain suppression is always required. The noise figure is of minor interest since the input power to the SOA is high.

To reach the nonlinear region of the SOA at reasonable input powers, a low saturation input power is desirable. A low saturation input power is achieved with a low modal cross-section, and a high differential gain. Thus, the IPDR values are small and, here, they are of less importance.

Additionally, fast gain and phase dynamics are important. A high bias current density is advantageous to decrease the effective carrier lifetime.

Finally, a high alpha-factor device is preferable for nonlinear operation due to the efficient coupling of gain and phase changes which can be used especially for cross-gain modulation (XGM) and cross-phase modulation (XPM) applications.

In summary, it is found that conventional SOA are preferable for nonlinear applications. Further, it should be noted that in QD SOA fast processes introduce a chirp, which is very small. Therefore, an efficient exploitation of this chirp in nonlinear interferometric schemes is quite challenging.

Wavelength Conversion with Nonlinear QD Semiconductor Optical Amplifiers

Wavelength converters are expected to play a crucial role in alleviating blocking issues in future high capacity transparent networks. A filter assisted wavelength conversion remains a challenging area for QD SOA. It is still unclear which should be the optimum scheme that takes into consideration the particular properties of the QD devices and ensures a high quality operation similar to conventional SOA.

The atom-like density of states and the existence of a carrier reservoir in QD SOA promise an efficient carrier refilling that may enable ultra-fast performance. In addition, the inhomogeneously broadened gain spectrum of QD SOA, originating from the size fluctuation of the QD, may open new opportunities for multi-wavelength operation. However, due to the low alpha factor in QD SOA, XPM as a nonlinear switching mechanism is eliminated, which leaves XGM as the dominant effect.

As a result in this thesis, filtering solutions for all-optical wavelength conversion with QD SOA were proposed and experimentally verified at 10, 40 and 80 Gbit/s, see *Chapter 5*. In a low carrier injection regime, QD SOA provide a limited amount of XPM and bit-pattern effects are observed. The optimum solution is a blue-shifted filtering scheme, implemented by two tunable delay interferometer (DI) and an optical band-pass filter (OBPF). The filter provides effective mitigation of the bit patterning by counterbalancing amplitude-to-amplitude and phase-to-amplitude conversions.

In a high carrier injection regime, bit patterning and phase effects do not exist. A wavelength converted signal suffering from a poor extinction ratio is produced. A notch filter that suppresses the signal's carrier frequency is considered as the remedy to this deficiency. A 1-bit DI can play this role producing an alternate mark inversion signal at its output.

In high-speed experiments, error-free (BER $< 10^{-9}$) 80 Gbit/s RZ OOK all-optical wavelength conversion from 1310 nm to 1320 nm using a QD SOA combined with a DI as subsequent filter was demonstrated, see *Section 5.3.2.3*. 40 Gbit/s wavelength conversion performed error-free without an error floor down to a BER of 10^{-12}. The fast gain dynamics of the QD resulted in open eye diagrams of the converted output signal of the QD SOA in front of the DI. Weak phase effects were still observed even in this moderate-carrier injection regime. Therefore, the slow carrier dynamics of the reservoir, which dominates the phase dynamics of the QD SOA, requires the application of the DI in a phase scheme, i.e., with large free spectral range values (320 GHz at 80 Gbit/s).

It also found in this work that the homogeneously broadened gain bandwidth is around 60 % of the total gain bandwidth which in consequence prevents multi-wavelength operation under gain saturation conditions.

Clock Recovery with Nonlinear QD Semiconductor Optical Amplifiers

An all-optical clock recovery functionality based on a Fabry-Pérot filter followed by a QD SOA at 40 Gbit/s was proposed and experimentally demonstrated, see *Section 5.4*. High quality clock extraction is achieved for single wavelength operation throughout the C-band for various pseudo-random binary sequences. The multi-wavelength performance was degraded due to induced cross-talk among the QD groups. The multi-wavelength performance of the circuit is expected to be significantly enhanced by properly engineering the QD.

Future Work on Nonlinear Semiconductor Optical Amplifiers

It is interesting to learn if a realization of multi-wavelength nonlinear signal processing applications using QD SOA is achievable using new device fabrication techniques. Possibly, the formation of a wetting layer in the growth of QD can be avoided, and tunnelling injection into the QD states can enable independent dot groups which then enable multi-wavelength operation.

In next-generation optical networks, the use of advanced modulation format data signals requires wavelength converters, clock recoveries, and regenerators that can cope with these formats. For this operation, it is interesting if SOA can also be used to demonstrate nonlinear applications such as phase-sensitive amplification based on cross-gain modulation, cross-phase modulation or four-wave mixing.

Realization of an All-Optical Grooming Switch

A novel all-optical switching node with grooming functionality and multi-wavelength regenerative capability was demonstrated, see *Chapter 6*. The node implementation demonstrated the high-quality interoperability of the optical time-division multiplexing to wavelength division multiplexing (OTDM-to-WDM), WDM-to-OTDM and all-optical 2R regeneration (re-amplification and re-shaping) subsystems for continuous traffic. Also, in a field trial using dark fiber links, the tolerance to impairments introduced by fiber transmission together with the switching of data between high bit rate 130 Gbit/s OOK metro-core rings and lower bit rate 43 Gbit/s OOK metro-access rings has been demonstrated with good performance. The node offers switching functionality in time, i.e., including time-slot interchanging, space, and wavelength domain.

A. Appendix

A.1. Modulation Formats: Transmitter

Modulators based on Lithium Niobate (LiNbO₃) are exclusively used in this thesis to encode OOK, PSK and 16QAM data signals onto the optical carrier. In this section, the intensity and amplitude transfer characteristics as a function of the voltage applied to a LiNbO₃ Mach-Zehnder interferometer (MZI) modulator is discussed. Additionally, an in-phase and quadrature-phase (IQ) modulator is introduced.

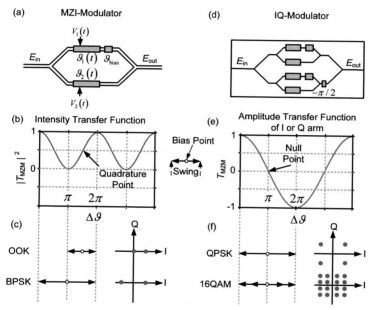

Fig.A. 1: Typical transmitter Mach-Zehnder interferometer (MZI) modulator and IQ modulator, transfer functions and OOK, PSK and QAM data encoding. (a) MZI schematic, (b) intensity-transfer function of the MZI, and (c) bias points and required swing (required phase shift of the MZI induced by a voltage) of the electrical data signal to generate an OOK or a BPSK data signal. (d) IQ modulator schematic, (e) amplitude-transfer function, and (f) bias points and electrical swing of the data signal in the electrical domain to generate QPSK and 16QAM data signals.

In MZI modulators, see Fig.A. 1(a), an optical continuous wave (cw) signal coupled into the device E_{in} is equally split onto two arms. A voltage $V_{1,2}(t)$ is applied on either one or both of the MZI arms. A phase change $\vartheta_{1,2}(t)$ can be impressed on an optical carrier wavelength by the electro-optic effect (Pockels effect). This phase modulation is converted into an amplitude or intensity modulation using the MZI. Additionally to the time-dependent drive signal, a DC voltage, the so-called bias V_{bias} is applied and induces a static phase shift ϑ_{bias}. Using this bias voltage the interferometer is set to the desired operating point [6].

The transfer function of the MZI modulator $T_{MZM}(t)$ for the input E_{in} and output field E_{out} can be written as

$$T_{MZM}(t) = \frac{E_{out}(t)}{E_{in}(t)} = \exp\left(j\frac{\vartheta_1(t)+\vartheta_2(t)}{2} + j\frac{\vartheta_{bias}}{2}\right)\cos\left(\frac{\vartheta_1(t)-\vartheta_2(t)}{2} + \frac{\vartheta_{bias}}{2}\right). \qquad (1.1)$$

Equation (1.1) shows, that there is usually both a phase modulation (from the first term) and an amplitude modulation (from the cosine term) passed on to a signal when going through a MZI modulator. The only exceptions are choices of pure phase modulation in the so-called push-push operation mode where $\vartheta_1(t) = \vartheta_2(t)$ or pure amplitude modulation in the so-called push-pull operation mode where $\vartheta_1(t) = -\vartheta_2(t)$. The push-pull operation mode is quite common and further discussed in the following [3].

The normalized intensity transfer function $|T_{MZM}|^2$ is depicted in Fig.A. 1(b) as a function of the total phase change $\Delta\vartheta(t) = 2\vartheta_1(t) + \vartheta_{bias}$ (including the phase change from the modulation and the bias) and can be written as

$$|T_{MZM}|^2 = \cos^2\left(\frac{\Delta\vartheta}{2}\right). \qquad (1.2)$$

If operated around the quadrature point (see Fig.A. 1(b)) the transfer function is quite linear in the intensity for small $\Delta\vartheta(t)$. A voltage required for a swing of $\Delta\vartheta = \pi$ is called V_π.

The normalized amplitude transfer function T_{MZM} is depicted in Fig.A. 1(e) as a function of the total phase change $\Delta\vartheta(t)$ and can be written as

$$T_{MZM} = \cos\left(\frac{\Delta\vartheta}{2}\right). \qquad (1.3)$$

If operated around the null point (see Fig.A. 1(e)) the transfer function is quite linear in the amplitude for small $\Delta\vartheta(t)$.

For intensity encoded (OOK) data signals, the bias voltage (corresponding phase change) is set to the quadrature point, see Fig.A. 1(b). The electrical binary (two-level) drive signal is applied with a total phase change of π (corresponding voltage swing). Since the MZI modulator is operated in the push-pull mode, an amplitude of $V_\pi / 2$ is applied for each arm. In the IQ constellation diagram the OOK shows one constellation point at the constellation "0" and a second constellation point at "+1". Typical ON/OFF extinction ratios are about 15 dB [6].

MZI modulators are also used for the generation of BPSK (DPSK) data signals in this thesis. First the bias voltage (corresponding phase) is set to the null point, see Fig.A. 1(b and e). Then, the electrical binary signal is applied with a total phase change of 2π (induced by voltage change). This way the electrical drive signal of each arm has to be set to V_π. In the IQ constellation diagram the BPSK shows one constellation point at "−1" and a second constellation point at "+1".

A common approach to generate QPSK and QAM data signals is by using an IQ modulator, see Fig.A. 1(d). An IQ modulator consists of an interleaved MZI with an MZI in the upper and lower arm. The two MZI are identical except for a phase-offset of $-\pi / 2$ in the lower arm with respect to the MZI in the upper arm. Such a modulator is arranged to modulate the amplitudes of the in-phase and the quadrature-phase of the complex amplitude separately. The MZI modulators in the lower and in the upper arm are operated in push-pull mode.

For QPSK data generation the bias of the MZI is set to the null point. The swing of the binary electrical data signals is set to a total phase change of 2π and both MZI modulators of the IQ modulator are operated with independent electrical data sources. Thus, four constellation points in the IQ plane are obtained, see Fig.A. 1(f).

An advantage of the IQ modulator is that it allows to universally address each and every point in the constellation diagram. There are several ways to generate a 16QAM signal. A most universal approach is the use of a software defined transmitter (multi-format transmitter) [157]. Here, two independent field-programmable arrays (FPGA) are used to generate digital electrical four-level signals for both the lower and the upper MZI. These signals are then converted by means of digital-to-analog converters (DAC) into four-level analog signals, see Fig.A. 1(f). These signals are applied to the upper and the lower arm of the MZM which both are operated in push-pull mode and at the null point. Other approaches are also discussed in the literature [133].

It should be mentioned that differential phase shift keying signals (DPSK, DQPSK) encode information on a phase difference between adjacent bits. If a pre-coding of the electrical data (drive) signals is used, it can be ensured that the received signals can be the original signals. Thus, the modulator setups known from BPSK and QPSK encoding can be applied to generate DPSK and DQPSK data signals. At DPSK modulation, it is sufficient to use a differential pre-coder with a one-bit delay feedback, whereas for DQPSK the pre-coding depends on the operation of the transmitter and receiver configurations [9].

In all experiments with differential phase-encoded data signals in this thesis, the pre-coding is not performed. However, to allow bit-error ratio measurements, the error detector is programmed with the expected data sequence. This is known due to the fact that pseudo-random bit sequences are used as test signals instead of real traffic.

Since all transmitters used for experiments in this thesis are operated in the push-pull mode, the instantaneous jumps in the phase are realized at the expense of some residual intensity modulation. Thus, amplitude (power) transitions occur in all experiments. To obtain a constant envelope PSK data signal, pure phase modulation is required. Here, a phase modulator can be used [3].

A.2. Delay Interferometer

A fundamental building block of differential (self-coherent) receivers and signal processing units used in this thesis is the delay interferometer (DI) filter shown in Fig.A. 2(a). The DI has two inputs $X_{i,j}$ and two outputs $Y_{i,j}$ and consists of two couplers with power coupling ratios of ideally 0.5 each. The waveguide, fiber or free space length between the couplers are L_1 and L_2.

The unit delay τ_D is set to the difference in path lengths $L_U = \Delta L = L_1 - L_2$. The magnitude of the frequency responses for each transfer function $|H_{i,j}|$ with $i,j = 1, 2$ are given by [230] assuming a lossless DI in which all ports are matched so that there are no reflections to be considered and the additional phase φ_D is set to zero (angular frequency ω)

$$|H_{i,j}| = |H_{j,i}| = \left|\frac{Y_i}{X_j}\right| = \left|\frac{Y_j}{X_i}\right|, \qquad |H_{i,j}|^2 = \cos^2\left(\frac{\omega\tau_D}{2}\right)$$

$$|H_{i,i}| = |H_{j,j}| = \left|\frac{Y_i}{X_i}\right| = \left|\frac{Y_j}{X_j}\right|, \qquad |H_{i,i}|^2 = \sin^2\left(\frac{\omega\tau_D}{2}\right). \tag{1.4}$$

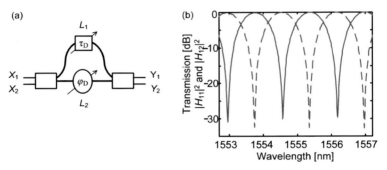

Fig.A. 2: Delay interferometer schematic and transfer function. (a) The DI has two inputs $X_{1,2}$ and two outputs $Y_{1,2}$, arm length L_1 and L_2, unit time delay τ_D and additional phase φ_D. (b) Power transfer functions at the destructive and constructive output ports.

Due to the loss and fabrication induced variations, the DI is not a lossless device and the power split ratio is not ideally 0.5. Thus, the ratio of the maximum to the minimum transmission over one free spectral range (FSR = $1/\tau_D = c_0 / (n_g L_U)$) is reduced to about 30 dB in the DI used for the experiments (n_g is the group refractive index). The corresponding power transfer functions are presented in Fig.A. 2(b).

The FSR of the DI is tunable via a free space delay. The mechanical tuning range amounts to a time delay tuning of 100 ps. Thus, FSR between 10 GHz to >> 400 GHz are achievable. The actuator used in the setup to shift the reflector has a mechanical step size of 2 nm [231].

A.3. Modulation Formats: Receiver

The design of an optical receiver depends largely on the modulation format used by the transmitter. In the following, typical receiver for intensity encoded (OOK), differentially-phase-encoded (DPSK, DQPSK), non-differentially phase-encoded (BPSK, QPSK) and QAM encoded data signals are presented.

An intensity-encoded data signal (OOK) can be directly detected with a pre-amplifier receiver using a fiber-amplifier (xDFA) with a very low noise figure (NF) and a high gain. The incoming data signal is filtered by an optical band-pass filter (OBPF), amplified, again filtered, converted into the electrical domain by a photodiode (PD), electronically amplified and analyzed with a bit-error ratio tester (BERT), see Fig.A. 3(a). The key elements are the PD and the optical amplifier. The PD demodulates the (optical) band-pass signal to an (electrical) baseband signal. Across a wide range of optical input powers, PD generate an electrical current that is linearly proportional to the optical intensity impinging on the diode, but it is inherently independent of the optical phase and also independent of the polarization [9]. The maximum signal-to-noise ratio (SNR) for a shot noise limited direct reception without optical pre-amplifier $SNR_{D,max}$ is ordinarily not achievable in an optical communications system due to the low receiver input power levels. Thus, an optical pre-amplifier is used, which reduces the achievable SNR ideally by a factor of 2. However in this configuration, one can approach the shot noise limited reception with realistic (low) receiver input power levels due to the high power levels at the PD which is guaranteed by the high optical pre-amplifier output power.

Non-differentially encoded PSK and M-ary QAM data signals can be detected with a receiver setup show in Fig.A. 3(b). Any modulation format in quadrature space that transports information using the optical phase, requires appropriate phase-to-intensity conversion prior to photodetection. Thus, the basic idea of coherent detection is to mix the received signal coherently with a phase-stable local oscillator (LO) before directing it onto the PD (coherent demodulation) [9]. If the received signal is in-phase with the LO, the PD records constructive interference, while it records destructive interference if the two optical fields have opposite phase, thus converting phase modulation into intensity-modulation amenable to photodetection.

A typical coherent receiver, according to Fig.A. 3(b), consists of a LO, a 90° optical hybrid, several PD, balanced detection, analog-to-digital (ADC) conversion and digital signal processing (DSP) units before the BERT analyzes the signal quality. The 90° optical hybrid is used to enable the extraction of the full information of the signal complex amplitude in the IQ-plane. Balanced detection is used here to subtract the dc parts from the LO and the signal, i.e., any LO relative intensity noise and the direct-detection signal component (without demodulation). It should be noted that, in contrast to delay demodulation, balanced detection does not impact the sensitivity of the coherent receiver in a fundamental way. Since the detected photocurrent (after the balanced detector) is proportional to the square root of the local oscillator power, the shot noise limit can be reached in coherent detection schemes by choosing the local oscillator power significantly high [9].

Three cases are distinguished: If the optical frequency of the LO f_{LO} and the signal carrier frequency f_s are precisely the same, an intermediate frequency of zero $f_{IF} = |f_s - f_{LO}| = 0$ is obtained. Here, robust phase-locking of the two optical signals is a challenge, which is currently successfully performed with DSP and highly stable LO. This case is called homodyne coherent reception. If the two frequencies are offset, the detected photocurrent shows oscillations at the intermediate frequency (IF). If the IF is chosen significantly higher than the signal bandwidth, the optical signal spectrum is downconverted into an electrical band-pass signal centered at the IF that can be further electronic processed. This case is called heterodyne coherent reception. If the IF is chosen to fall within the signal band, intradyne coherent reception is used [15].

The maximum achievable SNR for a shot noise limited reception with a homodyne coherent receiver is $SNR_{Hom} = 4\,SNR_{D,max}$, and with a heterodyne coherent receiver $SNR_{Het} = 2\,SNR_{D,max}$ (baseband signal) [232].

Typically to achieve a polarization independent coherent receiver two parallel coherent receiver setups, Fig.A. 3(b), are used, one for each of the two polarizations.

In this work, a homodyne coherent receiver is exclusively used for the reception of BPSK, QPSK and 16QAM data signals.

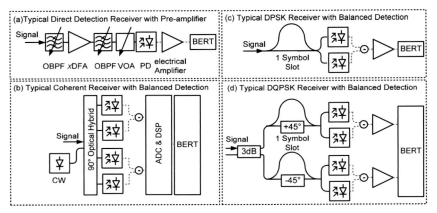

Fig.A. 3: Schematic of receivers for different modulation formats. (a) pre-amplifier receiver for direct detection of intensity-encoded signals, (b) typical coherent receiver to detect PSK and QAM data signals, (c) typical DPSK self-coherent (differential) receiver using balanced detection, and (d) typical DQPSK self-coherent receiver with balanced detection.

Differentially encoded PSK data signals are detected with receiver setups shown in Fig.A. 3(c) and (d). In a differential detector the signal is split onto two arms and recombined after one symbol duration. Thus, a phase-to-intensity conversion is performed before the PD. After recombining and properly setting the relative phases in the arms one obtains the constructive and destructive interference of a symbol with its previous symbol in the two outputs. Hence, the signal acts as a phase reference to itself and takes the role of the LO in a coherent receiver. This way differentially (quadrature) phase-encoded signals can be detected. If both signal contributions (I and Q) are to be detected, a second interferometer with 90° shifted interference properties is typically used, in analogy to a 90° hybrid in coherent detection.

It should be mentioned that the main advantage of DPSK compared to OOK comes from a 3 dB receiver sensitivity improvement if a balanced detection scheme is used.

In this work, the experimental investigation for DQPSK data signals is performed with a single DI in the receiver, but with two BER measurements with a DI at +45° and at −45°, respectively.

A.4. Multiplexing Techniques

Multiplexing is the art of combining N orthogonal symbols to allow for the simultaneous transmission in a shared channel (such as air, fiber, ...). Multiplexing techniques may exploit orthogonalities in space, time, frequency, polarization, quadrature, and code. In this thesis, the techniques in the time, quadrature and frequency are used.

In time-division multiplexing (TDM), multiple bit streams are multiplexed onto a single signal by assigning a recurrent time slot to each of the tributaries. Time pulses are short and spectrally broad, see Fig.A. 4(a). This makes TDM schemes challenging. The short pulses require a narrow receiver time window, and the large optical bandwidth makes precisely engineered dispersion compensation a necessity.

Alternatively, with wavelength-division-multiplexing (WDM), large aggregated bit rates are transmitted over optical fibers, see Fig.A. 4(b). In WDM, several wavelengths transport several data-streams in parallel. Pulse durations are longer and optical bandwidths are moderate. Spectral guard bands are required to avoid cross-talk from one channel to another.

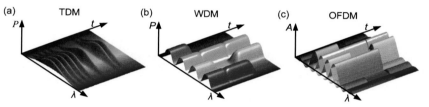

Fig.A. 4: Multiplexing techniques: (a) time-division multiplexing (TDM), (b) wavelength-division-multiplexing, and (c) orthogonal frequency-division multiplexing (OFDM). The figure is taken from [J3].

Orthogonal frequency-division multiplexing (OFDM) is a more recent approach in optical communications, although it is already well known in wireless data transmission. In contrast to WDM, modulated OFDM subcarriers overlap each other significantly, see Fig.A. 4(c). The encoding of information on a large number of subcarriers makes OFDM tolerant towards dispersion [J3]. In OFDM the information is transmitted in parallel on N subcarriers at a bit rate R_B/N instead of being transmitted in a serial stream at a high data rate on a single carrier. In the time domain, one can only see the overall envelope of the whole of the N subcarrier symbols. The OFDM signal actually resembles an analogue signal in time. The subcarrier frequencies are chosen such that they are spaced equidistantly. The symbol duration is the inverse of the subcarrier spacing (assuming no cyclic pre-fix etc.) [232].

The key building blocks of an (intensity-modulated) OFDM transmitter in DSP is a serial-to-parallel conversion of an incoming binary data sequence, the mapping of these data sequence onto the subcarriers (discrete frequency domain variables; mapping of the data onto constellation points of a modulation format), insertion of pilot tones for phase noise compensation at the Rx and training symbols, performing the inverse fast Fourier transform

(IFFT) to convert the modulated frequency domain variables into discrete time samples (summation over all subcarriers), parallel-to-serial conversion and insertion of cyclic pre-fix (CP) to enhance the dispersion tolerance (adding overhead). In the following, a digital-to-analog converter (in this thesis an arbitrary waveform generator (AWG)) is used to provide the required electrical input signal to the MZI modulator.

The key building blocks of an (intensity-modulated) OFDM receiver are the optical-to-electrical conversion with the PD and the sampling of the received OFDM data signal with a sampling scope (analog-to-digital conversion). Then DSP is used to remove the CP, to perform a serial-to-parallel conversion, to perform the FFT to obtain the discrete frequency domain samples from the discrete time samples, to remove the pilot tones, to perform the de-mapping of the signal, and to perform a parallel-to-serial conversion. Further details about OFDM can be found in [233].

In this thesis, a real valued OFDM signal is used. This is obtained by using a discrete multi-tone modulation (DMT). DMT is a sub-category of OFDM where the output of the IFFT is real. This means that only half of the FFT size is used for the modulation of the actual data.

A.5. Standard Single–Mode Fiber

In the experimental work, a standard single mode fiber (SMF) 28e from Corning Inc. is used. In the following some important fiber parameters are listed [234]:

- Core diameter: 8 µm
- Cladding diameter: 125 µm
- Numerical aperture: 0.14
- Zero dispersion wavelength: 1310 nm $< \lambda <$ 1324 nm
- Zero dispersion slope: 0.09 ps / (nm^2 km)
- Dispersion at 1550 nm: 18 ps / (km nm) and at 1625 nm: 22 ps/(km nm)
- Effective group index of refraction at 1310 nm: 1.4676 and at 1550 nm: 1.4682
- Attenuation at 1310 nm: 0.33-0.35 dB/km, at 1490 nm: 0.21-0.24 dB/km, at 1550 nm: 0.19-0.20 dB/km and at 1625 nm: 0.20-0.23 dB/km

A.6. Accuracy of Q^2 and EVM Method with Respect to BER

Accurate bit-error ratio (BER) estimations are of high importance for optical communication systems, e.g., to determine the signal quality after an SOA. Since direct BER measurements and simulations are very time consuming, it is common to estimate the BER based on analytic models for the probability density function (pdf) at the decision threshold [156], [235]. Evaluation methods such as the Q^2 factor and the EVM are often used to estimate the BER. The relation of the Q^2 factor and the EVM with the BER requires several assumptions for the pdf as already discussed in *Section 4.3.1* and *Section 4.4.1*.

Here, the accuracy of the used Q^2 factor and the EVM methods in relation to the real BER and in relation to the input power dynamic range (IPDR) is qualitatively discussed for the various modulation formats and the linear and nonlinear regime of SOA.

On-Off-Keying (OOK)

It is assumed that for low and moderate SOA input power levels amplified spontaneous emission (ASE) noise (added by the SOA) dominates at the receiver. Then, the statistics of the electrical noise at a direct-detection receiver are given for a bit-"1" (high-level) by a Rice-pdf and for a bit-"0" (low level) by a Rayleigh-pdf [236]. These statistics are however not used in practice due to the required knowledge of the complete pdf. Instead, a Gaussian model is used to describe the pdf. Then, the mean values and the variances of the mark and space levels (Q^2 factor method) are measured to estimate the BER. In many cases, the Q^2 factor method gives a reasonable estimation of the real BER [237]. Precise estimations have its limits once SOA nonlinearities (high SOA input powers) change significantly the pdf at the receiver [238].

In this thesis, the Q^2 method is used to evaluate the signal quality for OOK data signals (*Section 4.3*) at low and also at high SOA input powers. It is found that the measured Q^2 factor can be used to show BER trends and IPDR$_f$ performance estimations, irrespectively if low or high SOA input powers are applied. Comparing the SOA input power required to achieve a certain Q^2 factor limit and the corresponding BER limit, it is found that:

- At low SOA input power levels, the Q^2 factor and the BER limits are achieved approximately at identical SOA input powers. The input power variations are about ±0.5 dB (return-to-zero (RZ) and non-return-to-zero (NRZ)).
- At high SOA input power levels, the Q^2 factor gives conservative BER estimations. The SOA input power to achieve a given BER limit and the respective Q^2 limit can vary between +1...+4 dB for RZ and 0.5...+2 dB for NRZ modulation.
- The provided IPDR$_f$ in this thesis shows an accurate tendency for the different samples. However, it can be predicted that by a direct measurement of the BER to determine the input power dynamic range, even larger IPDR$_f$ values of 0.5...4 dB could be obtained.

Differentially-Phase-Shift-Keying (DPSK, DQPSK)

In the literature, the use of the Q^2 factor to estimate the BER for differentially-received PSK data signals is discussed controversially [239]-[241]. On one hand, it is shown that the usage of the Q^2 factor is not accurate [240]. On the other hand, it is shown that Gaussian pdf underestimates the tails of the pdf, but it can still be useful as a good approximation to the real pdf in the whole operation range of an SOA. This is because the major contributions to the differentially phase noise to the BER comes from the central part of the pdf where Gaussian approximation is quite accurate [241].

In this thesis, due to the discussed discrepancy, the BER is directly measured and the IPDR$_f$ is determined as a function of the power penalty at the receiver. It can be concluded that the error in the measurements of the IPDR$_f$ for 28 GBd DQPK (*Section 4.4.3*) is about ± 0.5 dB. However, the Q^2 method is applied in the simulations of the various differentially-phase-encoded data signals (*Section 4.4.1.3*) to evaluate the signal quality for low and also for high SOA input powers. Here, it can be concluded that the IPDR$_f$ results show the tendency for different alpha-factor SOA and the different modulation formats.

Quadrature-Amplitude-Modulation (QAM)

The EVM is a new method in optical communications. Recently, it has been found that the BER can be estimated from EVM data by an analytic relation (see *Section 4.4.1.2*). It is shown that the BER estimate is valid for data-aided reception and also nondata-aided reception if the received optical field is perturbed by Gaussian noise only. However, a study of the pdf and the corresponding relation of the EVM and the BER are still missing for SOA nonlinearities.

In laboratory experiments so far most receivers employ offline DSP at much reduced clock rates. This offline processing makes it very time consuming to reliably compute the BER, especially if the signal quality is high [156]. Especially, the measurement time for a BER of 10^{-9} for, e.g., a 20 GBd QPSK data signal is about half a day.

In this thesis, the EVM is used to determine the IPDR$_f$. At critical SOA input power levels, the BER has been also measured. It is found that for low SOA input power levels the measured BER corresponds to the estimated BER from the measured EVM. At high SOA input power levels, it has been found that the measured BER and the estimated BER are still comparable, but, the measured SOA input power level differences are in the range of $2\ldots3$ dB. Here, at high SOA input power levels, the EVM results show the tendency to underestimate the BER results. It can be concluded that the determined EVM method applied to QAM formats can predict the tendency of the IPDR$_f$. The measured IPDR$_f$ values using the EVM method are estimated to be $2\ldots3$ dB larger than the values obtained from a BER measurement. Further investigations of the relation between BER and EVM in the nonlinear regime of an SOA are required.

A.7. Frequency Resolved Electro–Absorption Gating

In this thesis, the frequency resolved electro-absorption gating (FREAG) method is used to evaluate the peak-to-peak phase changes after the SOA. Here, a brief explanation of this method is provided.

The FREAG method is a simple and highly sensitive pulse characterization technique in which a spectrogram of the train of pulses is measured using a temporal modulator as a gate [123] to determine the time-resolved amplitude and the phase of these pulses. In Fig.A. 5(a), a typical FREAG characterization setup is depicted. The transmitter (T_x) generates optical clock pulses with a repetition rate of $1/f_{Clock}$. The pulse train is amplified with an SOA and gated with an electro-absorption modulator (EAM). The gate has to be of comparable duration with the pulses under test. This is a significant advantage over sampling, where the temporal resolution is set by the duration of the gate. However, one of the prerequisites of the spectrogram technique is the presence of a clock signal synchronized with the periodic signal under test. The EAM gates (gating window $g(t)$) the pulse under test $E(t)$ in time and the Fabry-Pérot-filter (FPF, resolution of about 2 GHz) followed by the photodiode (PD) measures the power spectrum. Requirements on the speed of the rf electronics are small, as the period of the gate only needs to be a multiple of the period of the electric field under test. The power spectrum is measured for a large number of relative delays τ_{PG} between the gate and pulse under test, see Fig.A. 5(b). The delay is arranged with the phase shifter. Typically for a 100 ps gating window, 32 different time delays are measured. For each delay, typically a 500-point spectrum is acquired and it is resampled to 32 points, so that a 32×32 spectrogram is measured, see Fig.A. 5(c). This initial oversampling increases the SNR and provides immunity to small drifts of the setup [123].

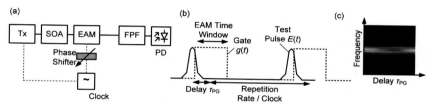

Fig.A. 5: Schematic of the frequency-resolved electro-absorption gating method, (a) setup, (b) functional principle, and (c) spectrogram.

From the time-frequency representation, the spectrogram, the complete information about the pulse and the gate is extracted without any assumptions. The optical pulse is represented by an analytic signal $E(t)$ and the pre-characterized EAM gate window has a response $g(t)$, both quantities being complex. The relative delay τ_{PG} can be set such that the output pulse of the EAM is related to the input pulse by $E'(t, \tau_{PG}) = E(t) g(t - \tau_{PG})$. One then builds the spectrogram $S(\omega, \tau_{PG})$ by measuring the spectrum of the gated pulse as a function of the optical angular frequency ω and relative delay τ_{PG}

$$S\left(\omega, \tau_{PG}\right) = \left| \int_{-\infty}^{+\infty} E'\left(t, \tau_{PG}\right) \exp\left(-j\omega t\right) dt \right|^2 . \tag{1.5}$$

The spectrogram is inverted with a principal component generalized projection algorithm using the power method [242]. This way the amplitude and the phase of the signal under test can be recovered.

In this thesis, a commercially available unit, EG 150 from Southern Photonics Ltd., is used to perform the FREAG measurements. General specifications of the unit are:
- Spectral resolution: 2 GHz
- Temporal resolution: with 32 point spectrum 3.1 ps, with 64 point spectrum 1.6 ps

A.8. Details on Experimental Setups

DQPSK Amplification Setup (Fig. 4.22)

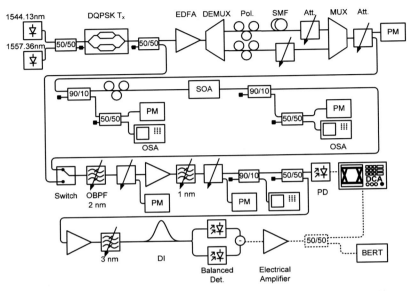

Fig.A. 6: Experimental setup for two channel 28 GBd DQPSK amplification measurements with an SOA. Tx: transmitter, EDFA: Erbium-doped fiber amplifier, DEMUX: demultiplexer, Pol.: polarization controller, SMF: 0.5 m of single-mode fiber for decorrelation, Att.: attenuator, MUX: multiplexer, PM: powermeter, OSA: optical spectrum analyzer, OBPF: optical band-pass filter, PD: photodiode, DCA: digital communications analyzer, DI: delay interferometer, BERT: bit-error ratio tester.

40 Gbit/s Wavelength Conversion Setup (results are presented in Fig. 5.11)

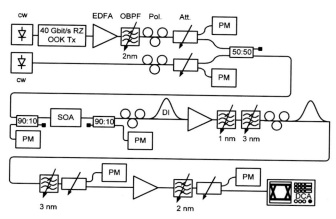

Fig.A. 7: Experimental setup for 40 Gbit/s RZ OOK wavelength conversion measurements with a QD SOA. cw: continuous wave laser, Tx: transmitter, EDFA: Erbium-doped fiber amplifier, OBPF: optical band-pass filter, Pol.: polarization controller, Att.: attenuator, PM: powermeter, DI: delay interferometer and DCA: digital communications analyzer.

Constants

$$h = 6.626\,068 \times 10^{-34}\,\text{Ws}^2 \tag{1.6}$$

$$e = 1.60217649 \times 10^{-19}\,\text{As} \tag{1.7}$$

$$\varepsilon_0 = 8.854\,188 \times 10^{-12}\,\frac{\text{As}}{\text{Vm}} \tag{1.8}$$

$$\mu_0 = 1.256\,637 \times 10^{-6}\,\frac{\text{Vs}}{\text{Am}} \tag{1.9}$$

$$c_0 = \frac{1}{\sqrt{\varepsilon_0 \mu_0}} = 299\,792\,458\,\frac{\text{m}}{\text{s}} \tag{1.10}$$

$$Z_0 = \sqrt{\frac{\mu_0}{\varepsilon_0}} = 376.7303\,\Omega \tag{1.11}$$

Glossary

Calligraphic Symbols

$\mathfrak{I}$ Linewidth broadening function

Greek Symbols

α_{CH}	Alpha factor for carrier heating (CH)
α_{Clad}	Cladding loss, unit m^{-1}
$\alpha_{Cloupling}$	Fiber-to-chip coupling loss, unit m^{-1}
α_{eff}	Time-dependent effective alpha factor
α_{int}	Internal material loss, unit m^{-1}
α_{H}	Alpha factor, linewidth-enhancement factor (Henry factor) for inter-band effects
α_{L}	Material loss, unit m^{-1}
α_{SHB}	Alpha factor for spectral-hole-burning (SHB)
β	Guided-wave propagation constant, unit m^{-1}
Γ	Confinement factor
γ_{inhom}	Inhomogeneous linewidth of QD ensemble (QD SOA model)
γ_{hom}	Homogeneous linewidth of resonant QD (QD SOA model)
ε_{0}	Electric permittivity of vacuum
ε_{c}	Nonlinear gain compression factor for inter-band effects
ε_{CH}	Nonlinear gain compression factor for carrier heating (CH)
$\bar{\varepsilon}_{r}$	Complex dielectric constant
ε_{r}	Real part of relative dielectric constant (scalar)
$-\varepsilon_{ri}$	Imaginary part of relative dielectric constant (scalar)
ε_{SHB}	Nonlinear gain compression factor for spectral-hole-burning (SHB)
η	Quantum efficiency
η_{i}	Injection efficiency of carriers
λ	Wavelength, unit m
λ_{dBr}	De Broglie wavelength
λ_{G}	Bandgap wavelength, unit m
λ_{Peak}	Gain peak wavelength, unit m
$\Delta\lambda$	Wavelength detuning, unit m
μ_{0}	Magnetic permeability of vacuum
$\mu_{m0,1}$	Mean value of "0" and "1" levels
$\Delta\nu_{FP}$	Fabry-Pérot-filter bandwidth
$\sigma_{0,1}$	Standard deviation of "0" and "1" levels
σ_{eff}	Effective cross-section for photon-carrier interaction

σ_i^2	Variance of photocurrent
σ_{res}	Resonant cross-section of the photon-carrier interaction
$\rho(k_{tr})$	Density of states (DOS) in the k_{tr}-space
ρ_{ASE}	Optical amplified spontaneous emission spectral density
$\rho_{ASE\parallel}$	Optical amplified spontaneous emission spectral density co-polarized with the signal
$\tau_{90\%-10\%}$	90 %-10 % gain recovery time, unit s
$\tau_{1,2}$	Characteristic time constants of carrier dynamics in QD SOA, unit s
τ_c	Effective carrier lifetime, unit s
$\tilde{\tau}_c$	Total effective carrier lifetime, unit s
τ_{Di}	Time delay in delay interferometer (DI) i
τ_{intra}	Intra-band carrier relaxation time
τ_p	Pulse width, unit s
τ_{ph}	Photon lifetime
$\tau_{s'}$	Lifetime for stimulated recombinations, unit s
ϕ	Optical phase
ϕ_e	Electrical phase
$\Delta\phi$	Optical phase change after SOA
φ_{Di}	Phase of delay interferometer (DI) i in reference to optical signal phase
χ	Susceptibility
ω	Angular frequency, unit rad/s
ω_s	Angular signal carrier frequency, unit rad/s

Latin Symbols

a	Differential gain
a_{La}	Lattice constant
$A(z,t)$	Slowly-varying envelope of a signal, unit $\sqrt{W}$
$A_{H,L}$	Field strength of H: logical "1" and L: logical "0"
B	Electrical bandwidth
B_G	3 dB gain bandwidth, unit m
B_O	Optical bandwidth
c_0	Speed of light in vacuum
c_p	Normalization constant
C	Area of active region
d	Active layer thickness
$D(W)$	Density of states (DOS) as a function of energy W
e	Elementary charge
E	Electric field, unit V/m
$E_{t,i}$	Transmitted reference constellation point vector
$E_{r,i}$	Received and measured signal vector
$E_{err,i}$	Error vector

$E_{t,m}$	Outermost constellation point vector
$f_{C,V}$	Fermi-function of conduction and valence band
f_e	Electrical modulation frequency (later identical to f_{rf} radio frequency)
f_i	Frequency with i: s signal carrier frequency; 1...n frequency of channel 1...n, unit Hz
f_{rf}	Radio-frequency
Δf_s	Chirp (instantaneous frequency shift)
F	Noise factor
$F(x,y)$	Transverse modal function
F_{ff}	Fiber-to-fiber noise factor
F_{total}	Total noise factor
g	Net modal gain m^{-1}
g_0	Net modal chip gain (unsaturated), unit m^{-1}
g_{di}	Degeneracy of energy level W_i
g_{lin}	Linear part of model net gain (QD SOA model)
g_m	Material gain, unit m^{-1}
g_{max}	Maximum gain of the resonant QD (QD SOA model)
g_{mL}	Net material gain, unit m^{-1}
g_{tot}	Total net modal gain including inter-band and intra-band effects
G	Chip gain
G	Gain rate
G_0	Small-signal (unsaturated) chip gain
G_{f0}	Small-signal (unsaturated) fiber-to-fiber gain
G_{ff}	Fiber-to-fiber gain
G_{Op}	SOA chip operation gain
G_{supp}	Suppressed gain
h	Planck constant, unit m
$h(t)$	Dimensionless gain coefficient for inter-band
$h_{tot}(t)$	Dimensionless gain coefficient
Δh_{CH}	Dimensionless gain compression coefficient for carrier heating (CH)
Δh_{SHB}	Dimensionless gain compression coefficient for spectral-hole-burning (SHB)
$\hbar$	Reduced Planck constant, $\hbar = 1.05457148 \times 10^{-34}$ m^{2}kg/s
$H(f_e)$	Optical small-signal transfer function
$H_m(f_e)$	Modulation response
H_{PROF}	Filter transfer function of pulse reformatting optical filter (PROF)
$\langle i \rangle$	Average photocurrent
I	Bias current, unit A
I_n	Number of received random symbols
J	Current density
j	Imaginary unit, $j^2 = -1$
k	Boltzmann constant

k_0	Free-space (vacuum) wavenumber, unit m^{-1}
k_M	Bits per symbol
k_{tr}	Electron-hole transition wave vector
l	Number of QD-layers
l_{fp}	Mean free path length, unit m
L	Length, unit m
m	Effective mass of charge carriers
$m_{o;in,out}$	Modulation depth at input and output
m_{OA}	Number of SOA
$\|M_{ave}\|^2$	Squared dipole matrix element
$\overline{n}$	Complex refractive index
n	Refractive index (real part)
$\overline{n}_b$	Complex background refractive index
n_b	Real part of complex background refractive index
n_{eff}	Real part of the effective refractive index
$-n_i$	Refractive index (imaginary part)
$\overline{n}_N$	Complex perturbation to the background refractive index
n_{sp}	Inversion factor
N_0	Carrier density for given I, τ_c, V
N_C	Normalized characteristic value of the carrier density (QD SOA model)
N_D	Density of QD per unit area in each layer
N_i	Total carrier density in energy state W_i, total number of microsystems in energy state W_i
N_J	Pumping strength (QD SOA model)
N_p	Photon number
N_t	Carrier density at transparency
N_{WL}	3D effective density of states for wetting layer (WL)
NF_{ff}	Fiber-to-fiber noise figure, unit dB
NF	Chip noise figure, unit dB
p_o	Occupation probability of the resonant QD (QD SOA model)
P	Power, unit W
P_{ASE}	SOA chip ASE output power
$P_{ASE\parallel}$	SOA chip ASE output power co-polarized with the signal
$P_{ASE,f}$	In-fiber amplified spontaneous emission output power, unit W
P_e	Electrical power, unit W
P_E	Electrical power (RoF section)
$P^{in,f}$	In-fiber input power, unit W
$P_{1,2}^{in,(f)}$	Lower (1) and upper (2) chip power limit of the IPDR (f, measured in the fiber)
$P^{out,f}$	In-fiber output power, unit W
P_{in}	Chip input power, unit W
P_O	Opitcal power (RoF section)

P_{out}	Chip output power, unit W
P_{s}	Saturation power, unit W
$P_{\text{sat}}^{\text{in,f}}$	In-fiber 3 dB saturation input power, unit W
$P_{\text{sat}}^{\text{in}}$	3 dB saturation input power
$P_{\text{sat}}^{\text{out,f}}$	In-fiber 3 dB saturation output power, unit W
$P_{\text{sat}}^{\text{out}}$	3 dB saturation output power
p_{tol}	Tolerance factor, linear
Δp_{tol}	Tolerance factor, unit dB
Q^2	Eye quality factor
r	Carrier generation rate
R	Responsivity
R_{ASE}	Recombination rate of carrier due to amplified spontaneous emission
R_{B}	Bit rate
R_{c}	Recombination rate of carriers due to radiative and nonradiative spontaneous transitions
R_{S}	Recombination rate of carriers due to signal amplification
R_{st}	Recombination rate of carriers due to stimulated transitions
R_{Sy}	Symbol-rate
S	Signal photon density
S'	Total photon density of signal and amplified spontaneous emission noise
S_{ASE}	Amplified spontaneous emission noise photon density
S_{s}	Saturation photon density
SNR	Signal-to-noise ratio
t	Time, unit s
t_{wg}	Effective optical thickness of the QD SOA waveguide (QD SOA model)
T	Temperature, unit K
$u_{1,0}$	Mean value of "0" and "1" levels
U	Voltage, V
v_g	Group velocity, unit m/s
V	Volume of active region
w	Waveguide width
$w_{\text{C,V}}$	Band energies of the conduction band and of the valence band
$w_{\text{p}}(W_i)$	Occupation probability of various energy levels W_i
$W_0(N)$	QD filling factor (QD SOA model)
$W_{\text{C,V}}$	Conduction band and valence band edge energy
W_{F}	Fermi energy
$W_{\text{Fn,Fp}}$	Quasi-Fermi levels for electrons in the conduction band and holes in the valence band
W_{G}	Bandgap energy
W_i	Energy states (sublevels) in semiconductor; microsystem
x	Material fraction
x	(Horizontal) spatial coordinate, unit m

y	Material fraction
y	(Vertical) spatial coordinate, unit m
z	Material fraction
z	(Longitudinal) spatial coordinate, unit m
Z_0	Free-space wave impedance

Acronyms

3R	Reamplification, reshaping, retiming
ADM	Add-drop multiplexer
ADORE	Asynchronous digital optical regenerator
AM	Amplitude modulation
AMI	Alternate mark inversion (format)
AOM	Acousto-optic modulator
ASE	Amplified spontaneous emission
ASK	Amplitude-shift keying (format)
Att.	Attenuator
Au	Auger (recombination)
BER	Bit-error ratio
BERT	Bit-error ratio tester
BSOF	Blue-shifted optical filter
CB	Conduction band
CC	Carrier cooling
CH	Carrier heating
CO	Central office
CRU	Clock-recovery unit
cw	Continuous-wave
CWDM	Coarse wavelength-division-multiplexing
DCA	Digital communications analyzer
DEMUX	Demultiplexer
DI	Delay interferometer
DOS	Density of states
DQPSK	Differential quadrature phase-shift keying (format)
DSL	Digital-subscriber line
DSP	Digital signal processing
DUT	Device under test
DWDM	Dense wavelength-division-multiplexing
DXC	Digital-cross connect
EAM	Electro-absorption modulator
EDFA	Erbium-doped fiber amplifier

EPON	Ethernet PON, ethernet passive optical network
ER	Extinction ratio
ESA	Electrical spectrum analyzer
ETDM	Electrical time-division multiplexing
FCA	Free-carrier absorption
FEC	Forward error correction
FPF	Fabry-Pérot filter
FREAG	Frequency-resolved electro-absorption gating (technique)
FSR	Free-spectral range
FtF	Fiber-to-Fiber
FTTH	Fiber-to-the-Home
FWHM	Full width at half maximum
FWM	Four-wave mixing
GPON	Gigabit PON, gigabit passive optical network
HDTV	High-definition television
HNLF	Highly-nonlinear fiber
IP	Internet protocol
IPDR	Input power dynamic range (optical)
LCA	Lightwave component analyzer
MAC	Media access control (protocol)
MAN	Metropolitan-area network
MBE	Molecular beam epitaxy
MLL	Mode-locked laser
MOCVD	Metal organic chemical vapour deposition
MZI	Mach-Zehnder interferometer
NF	Noise figure
NLSE	Nonlinear Schrödinger equation
NRZ DQPSK	Non-return-to-zero differential quadrature phase-shift keying (format)
NRZ OOK	Non-return-to-zero on-off keying (format)
OADM	Optical add-drop multiplexer
OBPF	Optical band-pass filter
OEO	Optical-to-electrical-to-optical (conversion)
OFDM	Optical orthogonal frequency-division multiplexing
OLT	Optical line terminal, sometimes also used with: termination
ONT	Optical-network termination
OPO	Optical parametric oscillator
OSA	Optical spectrum analyzer
OSNR	Optical signal to noise ratio
OTDM	Optical time-division multiplexing
OXC	Optical-cross connect
PD	Photodiode
PIC	Photonic integrated circuit

PON	Passive optical network
PDFA	Praseodymium-doped fiber amplifier
PDG	Polarization dependent gain
PLL	Phase-locked loop
PM	Power meter
PMF	Polarization-maintaining fiber
Pol.	Polarization controller
PRBS	Pseudo-random bit sequence
PROF	Pulse reformatting optical filter
PSK	Phase-shift keying (format)
PtP	Peak-to-Peak
QAM	Quadrature-amplitude modulation
QD	Quantum dot
QDash	Quantum dash
QD SOA	Quantum-dot semiconductor optical amplifier
QW	Quantum well
RE	Reach extender
rms	root mean square
ROADM	Reconfigurable optical add-drop multiplexer
RoF	Radio-over-Fiber
RSOA	Reflective-semiconductor optical amplifier
RSOF	Red-shifted optical filter
Rx	Receiver
RZ OOK	Return-to-zero on-off keying (format)
SHB	Spectral-hole burning
SK	Stranski-Krastanov (growth mode)
SMF	Single-mode fiber
SNR	Signal-to-noise ratio
SOA	Semiconductor optical amplifier
sp	spontaneous (emission)
SPM	Self-phase modulation
SRH	Shockly-Rhead-Hall (recombination)
SVEA	Slowly-varying envelope approximation
TEC	Temperature controller
TE	Transverse electric (mode)
TLS	Tunable laser source
TM	Transverse magnetic (mode)
TMLL	Tunable mode-locked laser
TPA	Two-photon absorption
TSI	Time-slot interchanging
Tx	Transmitter
UMI	Unbalanced Michelson interferometer

UNI	Ultra-fast nonlinear interferometer
VB	Valence band
VOA	Variable optical attenuator
VoD	Video-on-demand
VoIP	Voice-over-IP
WAN	Wide-area network
WDM	Wavelength-division-multiplexing
WL	Wetting layer
WSS	Wavelength-selective switch
XGM	Cross-gain modulation
XPM	Cross-phase modulation

References

[1] "Cisco visual networking index: Forecast and methodology, 2010-2015," Cisco Systems, Inc., San Jose, CA, USA, white paper (2009) [online]

[2] M. Sexton and A. Reid, *Transmission Networking: SONET and the Synchronous Digital Hierarchy* (Artech House, Boston, 1992)

[3] J. Leuthold, *Optische Kommunikationssysteme* (Institute of Photonics and Quantum Electronics, Karlsruhe, Germany, 2009), Lecture notes

[4] J. Prat, *Next-generation FTTH passive optical networks: research towards unlimited bandwidth access* (Springer, 2008)

[5] B. Mukherjee, *Broadband access networks* (Springer, Dordrecht, 2009)

[6] P. Vorreau, *An optical grooming switch for high-speed traffic aggregation*, PhD thesis, University of Karlsruhe (TH) (2010)

[7] "Entering the Zettabyte Era," Cisco Systems, Inc., San Jose, CA, USA, white paper (2009) [online]

[8] T. Koonen, "Fiber to the home/fiber to the premises: what, where and when?," Proc. IEEE **94**, 911-934 (2006)

[9] I. Kaminow, T. Li, and A. E. Willner, *Optical fiber telecommunications V*, B: systems and networks (Elsevier, Amsterdam, 2008)

[10] R. P. Davey, D. B. Grossman, M. Rasztovits-Wiech, D. B. Payne, D. Nesset, A. E. Kelly, S. Appathurai, and S.-H. Yang, "Long-Reach Passive Optical Networks," J. Lightwave Technol. **27**, 273-291 (2009)

[11] N. Cvijetic, "OFDM for next-generation optical access networks," J. Lightwave Technol. **30**, 384-398 (2012)

[12] Gigabit-capable Passive Optical Networks (GPON): Physical Media Dependent (PMD) layer specification, ITU-T Recommendation G.984.2, 03/2003 (2003)

[13] H. G. Krimmel, T. Pfeiffer, B. Deppisch, and L. Jentsch, "Hybrid electro-optical feedback gain-stabilized EDFAs for long-reach wavelength-multiplexed passive optical networks," *35th European Conference on Optical Communication* (ECOC '09), 2009, paper 9.5.3

[14] H. Rohde, S. Smolorz, S. Wey, and E. Gottwald, "Coherent Optical Access Networks," in *Optical Fiber Communication Conference*, OSA Technical Digest (CD) (Optical Society of America, 2011), paper OTuB1

[15] P. J. Winzer, and R.-J. Essiambre, "Advanced modulation formats for high-capacity optical transport networks," J. Lightwave Technol. **24**, 4711-4728 (2006)

[16] P. J. Winzer, and R.-J. Essiambre, "Advanced optical modulation formats," Proc. IEEE **94**, 952-985 (2006)

[17] D. R. Zimmerman and L. H.Spiekman, "Amplifiers for the masses: EDFA, EDWA, and SOA amplets for metro and access applications," J. Lightwave Technol. **22**, 63-70 (2004)

[18] K. E. Stubkjaer, "Semiconductor Optical Amplifier-Based All-Optical Gates for High-Speed Optical Processing," IEEE J. Select. Topics Quantum Electron. **6**, 1428-1435 (2000)

[19] L. A. Coldren, S. C. Nicholes, L. Johanssion, S. Ristic, R. S. Guzzon, E. J. Norberg, and U. Krishnamachari, "High performance InP-Based Photonic ICs - A Tutorial," J. Lightwave Technol. **29**, 554-570 (2011)

[20] N. K. Dutta and Q. Wang, *Semiconductor Optical Amplifiers* (World Scientific Publishing, Singapore, 2006)

[21] K. Morito, "High-power semiconductor optical amplifier," *Opt. Fiber Commun. Conf. and Nat. Fiber Opt. Eng. Conf.* (OFC/NFOEC'09), Techn. Digest (San Diego, CA, USA, 2009) paper OWQ4 (tutorial)

[22] C. Michie, A. E. Kelly, J. McGeough, I. Armstrong, I. Andonovic, and C. Tombling, "Polarization-insensitive SOA using strained bulk active regions," J. Lightwave Technol. **24**, 3920-3927 (2006)

[23] M. Sugawara, H. Ebe, N. Hatori, and M. Ishida, Y. Arakawa, T. Akiyama, K. Otsubo, and Y. Nakata, "Theory of optical signal amplification and processing by quantum-dot semiconductor optical amplifiers," Phys. Rev. B **69**, 235332 (2004)

[24] R. Brenot, F. Lelarge, O. Legouezigou, F. Pommereau, F. Poingt, L. Legouezigou, E. Derouin, O. Drisse, B. Rousseau, F. Martin, and G. H. Duan, "Quantum dots semiconductor optical amplifier with a -3 dB bandwidth of up to 120 nm in semi-cooled operation," in *Opt. Fiber Commun. Conf. and Nat. Fiber Opt. Eng. Conf.* (OFC/NFOEC'08), Techn. Digest (San Diego, CA, USA, 2008) paper OTuC1

[25] K. Morito, M. Ekawa, T. Watanabe, and Y. Kotaki, "High-output-power polarization-insensitive semiconductor optical amplifier," J. Lightwave Technol. **21**, 176-181 (2003)

[26] M. G. A. Bernard and G. Duraffourg, "Laser conditions in semiconductors amplifier," phys. stat. sol. **1**, 699 (1961)

[27] N. Nakamura and S. Tsuji, "Single-mode semiconductor injection lasers for optical fiber communications," IEEE J. Quantum Electron. **QE-17**, 994 (1981)

[28] G. Grau and W. Freude, *Optische Nachrichtentechnik* (Springer, Berlin, 1991)

[29] Z. I. Alferov, "Nobel Lecture: The double heterostructure concept and its applications in physics, electronics, and technology," Rev. Mod. Phys. **73**, 767-782 (2001)

[30] G. Lasher and F. Stern, "Spontaneous and stimulated recombination radiation in semiconductors," Phys. Rev. **133**, A553 (1964)

[31] P. S. Zory, Jr., *Quantum Well Laser* (Academic Press, San Diego, 1993)

[32] J. Leuthold, M. Mayer, J. Eckner, G. Guekos, H. Melchior, and Ch. Zellweger, "Material gain of bulk 1.55 μm InGaAsP/InP semiconductor optical amplifiers approximated by a polynomial model," J. Appl. Phys. **87**, 618-620 (2000)

[33] J. Leuthold, *Advanced indium-phosphide waveguide Mach-Zehnder interferometer all-optical switches and wavelength converters*, Series in Quantum Electronics, vol. 12 (Hartung-Gorre, Konstanz, Germany, 1999)

[34] D. Bimberg, M. Grundmann, and N. N. Ledentsov, *Quantum Dot Heterostructures* (John Wiley & Sons, Chichester, 1999)

[35] M. A. Newkirk, B. I. Miller, U. Koren, M. G. Young, M. Chien, R. M. Jopson, and C. A. Burrus, "1.5 μm multiquantum-well semiconductor optical amplifier with tensile and compressively strained wells for polarization-independent gain," IEEE Photon. Technol. Lett. **4**, 406-408 (1993)

[36] K. Magari, M. Okamoto, Y. Suzuki, K. Sato, Y. Noguchi, and O Mikami, "Polarization-insensitive optical amplifier with tensile-strained-barrier MQW structure," IEEE J. Quantum Electron. **30**, 695-702 (1994)

[37] D. Leclerc, P. Brosson, F. Pommereau, R. Ngo, P. Doussière, F. Mallécot, P. Gavignet, I. Wamsler, G. Laube, W. Hunziker, W. Vogt, and H. Melchior, "High-performance semiconductor optical amplifier array for self-aligned packaging using Si *V*-groove flip-chip technique," IEEE Photon. Technol. Lett. **7**, 476-478 (1995)

[38] C.-E. Zah, R. Bhat, B. N. Pathak, F. Favire, W. Lin, M. C. Wang, N. C. Andreadakis, D. M. Hwang, M. A. Koza, T.-P. Lee, Z. Wang, D. Darby, D. Flanders, and J. J. Hsieh, "High-performance uncooled 1.3-μm $Al_xGa_yIn_{1-x-y}As$/InP strained-layer quantum-well lasers for subscriber loop applications," IEEE J. Quantum Electron. **30,** 511-523 (1994)

[39] M. Yamada, T. Anan, K. Tokutome, and S. Sugou, "High-temperature characteristics of 1.3-μm InAsP-InAlGaAs ridge waveguide lasers," IEEE Photon. Technol. Lett. **11**, 164-166 (1999)

[40] P. Koonath, S. Kim, W.-J. Cho, and A. Gopinath, "Polarization-insensitive optical amplifiers in AlInGaAs," IEEE Photon. Technol. Lett. **13**, 779-781 (2001)

[41] H. Ma, X. Yi, and S. Chen, "1.55 μm AlGaInAs/InP polarization-insensitive optical amplifier with tensile strained wells grown by MOCVD," Opt. and Quantum Electron. **35**, 1107-1112 (2003)

[42]   J. Hashimoto, K. Koyama, T. Katsuyama, Y. Tsuji, K. Fujii, K. Yamazaki, and A. Ishida, "1.3 µm GaInNAs bandgap difference confinement semiconductor optical amplifiers," Jpn. J. Appl. Phys. **45**, 11635-1639 (2006)

[43]   S. Tanaka, A. Uetake, S. Yamazaki, M. Ekawa, and K. Morito, "Polarization-insensitive GaInNAs–GaInAs MQW-SOA with low noise figure and small gain tilt over 90-nm bandwidth (1510–1600 nm)," IEEE Photon. Technol. Lett. **20**, 1311-1313 (2008)

[44]   S. J. Sweeney, patent WO 2010/149978 A1

[45]   S. J. Sweeney, "Bismide-alloys for higher efficiency infrared semiconductor lasers," *22nd IEEE Int. Semicond. Laser Conf.* (ISLC2010), Conf. Digest (Kyoto, Japan, 2010) paper P24

[46]   Y. Tominaga, K. Oe, and M. Yoshimoto, "Low temperature dependance of oscillation wavelength in GaAs$_{1-x}$Bi$_x$ laser by photo-pumping," Appl. Phys. Express **3**, 062201 (2010)

[47]   M. Yoshimoto, W. Huang, G. Feng, and K. Oe, "New semiconductor alloy GaNAsBi with temperature-insensitive bandgap," phys. stat. sol. (b) **243**, 1421–1425 (2006)

[48]   J. P. Reithmaier, G. Eisenstein, and A. Forchel, "InAs/InP quantum-dash lasers and amplifiers," Proc. IEEE **95**, 1779-1790 (2007)

[49]   F. Lelarge, B. Dagens, J. Renaudier, R. Brenot, A. Accard, F. van Dijk, D. Make, O. L. Gouezigou, J.-G. Provost, F. Poingt, J. Landreau, O. Drisse, E. Derouin, B. Rousseau, F. Pommereau, and G.-H. Duan, "Recent advances on InAs/InP quantum dash based semiconductor lasers and optical amplifiers operating at 1.55 µm," IEEE J. Select. Topics Quantum Electron. **13**, 111-124 (2007)

[50]   T. Akiyama, M. Sugawara, and Y. Arakawa, "Quantum-dot semiconductor optical amplifiers," Proc. IEEE **95**, 1757-1766 (2007)

[51]   A. R. Kovsh, N. A. Maleev, A. E. Zhukov. S. S. Mikhrin, A. P. Vasil'ev, E. A. Semenova, Y. M. Shernyakov, M. V. Maximov, D.A. Livshits, V. M. Ustinov. N. N. Ledentsov, D. Bimberg, and Z. I. Alferov, "InAs/InGaAs/GaAs quantum dot lasers of 1.3 µm range with enhanced optical gain," J. Cryst. Growth **251**, 729-736 (2003)

[52]   D. Bimberg, G. Fiol, M. Kuntz, C. Meuer, M. Laemmlin, N. N. Ledentsov, and A. R. Kovsh, "High speed nanophotonic devices based on quantum dots," phys. stat. sol. (a) **203**, 3523-3532 (2006)

[53]   T. Kita, O. Wada, H. Ebe, Y. Nakata, and M. Sugawara, "Polarization-independent photoluminescence from columnar InAs/GaAs self-assembled quantum dots," Jpn. J. Appl. Phys. **41**, L1143-L1145 (2002)

[54] N. Yasuoka, K. Kawaguchi, H. Ebe, T. Akiyama, M. Ekawa, K. Morito, M. Sugawara, Y. Arakawa, "1.55-μm polarization-insensitive quantum dot semiconductor optical amplifier," _Europ. Conf. Opt. Commun._ (ECOC'08), Tech. Digest (Brussels, Belgium, 2008), paper Th.1.C.1

[55] N. Yasuoka, K. Kawaguchi, H. Ebe, T. Akiyama, M. Ekawa, K. Morito, M. Sugawara, and Y. Arakawa, "Quantum-dot semiconductor optical amplifiers with polarization-independent gains in 1.5-μm wavelength bands," IEEE Photon. Technol. Lett. **20**, 1908-1910 (2008)

[56] D. Litvinov, H. Blank, D. Schneider, D. Gerthsen, T. Vallaitis, J. Leuthold, T. Passow, A. Grau, H. Kalt, C. Klingshirn, and M. Hetterich, "Influence of InGaAs cap layers with different in concentration on the properties of InGaAs quantum dots," J. Appl. Phys. **103**, 083532 (2008)

[57] H. Wang, E. Aw, M. Xia, M. Thompson, R. Penty, I. White, and A. Kovsh, "Temperature independent optical amplification in uncooled quantum dot optical amplifiers," in _Opt. Fiber Commun. Conf. and Nat. Fiber Opt. Eng. Conf._ (OFC/NFOEC'08), Techn. Digest (San Diego, CA, USA, 2008) paper OTuC2

[58] R. Brenot, M. D. Manzanedo, J.-G. Provost, O. Legouezigou, F. Pommereau, F. Poingt, L. Legouezigou, E. Derouin, O. Drisse, B. Rousseau, F. Martin, F. Lelarge, and G. H. Duan, "Chirp reduction in quantum dot-like semiconductor optical amplifiers," _Europ. Conf. Opt. Commun._ (ECOC'07) (Berlin, Germany, 2007), paper We08.6.6

[59] M. Spyropoulou, S. Sygletos, and I. Tomkos, "Simulation of multi-wavelength regeneration based on QD semiconductor optical amplifiers," IEEE Photon. Technol. Lett. **19**, 1577-1579 (2007)

[60] G. P. Agrawal and N. K. Dutta, _Semiconductor Lasers_, 2nd ed. (Van Nostrand Reinhold, New York, 1993)

[61] G. P. Agrawal, _Fiber-Optic Communication Systems_, (Wiley & Sons, New York, 2002)

[62] M. J. Connelly, _Semiconductor Optical Amplifiers_, (Kluwer Academic Publishers, Boston, 2002)

[63] "Photonic integrated circuits-A technology and applications primer," Infinera, white paper, www.infinera.com (2005)

[64] R. Nagarajan, M. Kato, J. Pleumeekers, P. Evans, D. Lambert, A. Chen, V. Dominic, A. Mathur, P. Chavarkar, M. Missey, A. Dentai, S. Hurtt, J. Bäck, R. Muthiah, S. Murthy, R. Salvatore, C. Joyner, J. Rossi, R. Schneider, M. Ziari, H.-S. Tsai, J. Bostak, M. Kaufmann, S. Pennypacker, T. Butrie, M. Reffle, D. Mehuys, M. Mitchell, A. Nilsson, S. Grubb, F. Kish, and D. Welch, "Large-scale photonic integrated circuits for long-haul transmission and switching," J. of Opt. Netw. **6**, 102-111 (2007)

[65]   T. Durhuus, B. Mikkelsen, C. Joergensen, S. L. Danielsen, and K. E. Stubkjaer, "All-Optical Wavelength Conversion by Semiconductor Optical Amplifiers," J. Lightwave Technol. **14**, 942-954 (1996)

[66]   J. Leuthold, C. H. Joyner, B. Mikkelsen, G. Raybon, J. L. Pleumeekers, B. I. Miller, K. Dreyer, C. A. Burrus; "100 Gbit/s all-optical wavelength conversion with integrated SOA delayed-interference configuration," Electron. Letters **36**, 1129-1130 (2000)

[67]   Y. Liu, E. Tangdiongga, Z. Li, H. de Waardt, A. M. J. Koonen, D. Khoe, H. J. S. Dorren, X. Shu, and I. Bennion, "Error-free 320 Gb/s SOA-based wavelength conversion using optical filtering," in *Opt. Fiber Commun. Conf. and Nat. Fiber Opt. Eng. Conf.* (OFC/NFOEC'06), Techn. Digest, paper PDP28

[68]   M. Matsuura, O. Raz, F. Gomez-Agis, N. Calabretta, and H. J. S. Dorren, "320-Gb/s wavelength conversion based on cross-gain modulation in a quantum-dot SOA," *Europ. Conf. Opt. Commun.* (ECOC'11) (Geneva, Switzerland, 2011), paper Mo.1.A.1

[69]   J. Leuthold, L. Möller, J. Jaques, S. Cabot, L. Zhang, P. Bernasconi, M. Cappuzzo, L. Gomez, E. Laskowski, E. Chen, A. Wong-Foy, and A. Griffin, "160 Gbit/s SOA all-optical wavelength converter and assessment of its regenerative properties," Electronics Letters **40**, 554-555 (2004)

[70]   M. L. Nielsen and J. Mørk, "Increasing the modulation bandwidth of semiconductor-optical-amplifier-based switches by using optical filtering," J. Opt. Soc. Am. **B 21**, 1606-1619 (2004)

[71]   J. Leuthold, D. M. Marom, S. Cabot, J. J. Jaques, R. Ryf, and C. Randy, "All-optical wavelength conversion using a pulse reformatting optical filter," J. Lightwave Technol. **22**, 186-192 (2004)

[72]   Y. Liu, E. Tangdiongga, Z. Li, S. Zhang, H. de Waardt, G. D. Khoe, and H. J. S. Dorren, "Error-Free All-optical wavelength conversion at 160 Gb/s using a Semiconductor Optical Amplifier and an optical bandpass filter," J. Lightwave Technol. **24**, pp. 230-236 (2006)

[73]   J. Leuthold, R. Ryf, D. N. Maywar, S. Cabot, J. Jaques, and S. S. Patel, "Nonblocking all-optical cross connect based on regenerative all-optical wavelength converter in a transparent demonstration over 42 nodes and 16800 km," J. Lightwave Technol. **21**, 2863-2869 (2003)

[74]   C. Schubert, *Interferometric Gates for All-Optical Signal Processing*, PhD thesis, University Berlin (TH) (2004)

[75]   K. N. Nguyen, T. Kise, J. M. Garcia, H. N. Poulsen, and D. J. Blumenthal, "All-optical 2R regeneration of BPSK and QPSK data using a 90 optical hybrid and integrated SOA-MZI wavelength converter pairs," in *Opt. Fiber Commun. Conf. and Nat. Fiber Opt. Eng. Conf.* (OFC/NFOEC'11), Techn. Digest, paper OMT3

[76] M. Bougioukos, T. Richter, C. Kouloumentas, V. Katopodis, R. Harmon, D. Rogers, J. Harrison, A. Poustie, G. Maxwell, C. Schubert, and H. Avramopoulos, "Phase-incoherent DQPSK wavelength conversion using a photonic integrated circuit," IEEE Photon. Technol. Lett. **23**, 1649-1651 (2011)

[77] M. Spyropoulou, M. Bougioukos, G. Giannoulis, Ch. Kouloumentas, D. Kalavrouziotis, A. Maziotis, P. Bakapoulos, R. Harmon, D. Rogers, J. Harrison, A. Poustie, G. Maxwell, and H. Avramopoulos, "Large-scale photonic integrated circuit for multi-format regeneration and wavelength conversion," in *Opt. Fiber Commun. Conf. and Nat. Fiber Opt. Eng. Conf.* (OFC/NFOEC'11), Techn. Digest, paper OThY2

[78] R. Slavik, F. Parmigiani, J. Kakande, C. Lundström, M. Sjödin, P. A. Andrekson, R. Weerasuriya, S. Sygleots, A. D. Ellis, L. Grüner-Nielsen, D. Jakobson, S. Herstrøm, R. Phelan, J. O'Gorman, A. Bogris, D. Syvridis, S. Dasgupta, P. Petropoulos, and D. J. Richardson, "All-optical phase and amplitude regenerator for next-generation telecommunications systems," Nature Photonics **4**, 690-695 (2010)

[79] J. Leuthold, C. Koos, and W. Freude, "Nonlinear silicon photonics," Nature Photonics **4**, 535-544 (2010)

[80] F. Parmigiani, P. Vorreau, L. Provost, K. Mukasa, M. Takahashi, M. Tadakuma, P. Petropoulos, D. J. Richardson,W. Freude, and J. Leuthold, "2R regeneration of two 130 Gbit/s channels within a single fiber," in *Opt. Fiber Commun. Conf. and Nat. Fiber Opt. Eng. Conf.* (OFC/NFOEC'09), Techn. Digest, paper JThA56

[81] T. Kremp, *Split-step wavelet collocation methods for linear and nonlinear optical wave propagation*, PhD thesis, University of Karlsruhe (TH) (2002)

[82] J. Wang, *Pattern effect mitigation techniques for all-optical wavelength converters based on semiconductor optical amplifiers*, PhD thesis, University of Karlsruhe (TH) (2008)

[83] H. A. Kramers, "Diffusion of light by atoms," Atti. Congr. Internat. Fisici, **2**, 545-57 (1927)

[84] R. de L. Kronig, "On the theory of the dispersion of X-rays," J. Opt. Soc. Am. **12**, 547-557 (1926)

[85] D.C. Hutchings, M. Sheik-Bahae, D.J. Hagan, and E.W. van Stryland, "Kramers-Krönig relations in nonlinear optics," Opt. and Quantum Electron. **24**, 1-30 (1992)

[86] C.H. Henry, "Theory of the linewidth of semiconductor lasers," IEEE J. Quantum Electron. **QE-18**, 259-264 (1982)

[87] L. Occhi, L. Schares, and G. Guekos, "Phase modeling based on the α-factor in bulk semiconductor optical amplifiers," IEEE J. Select. Topics Quantum Electron. **9**, 788-797 (2003)

[88] G. Agrawal, *Nonlinear Fiber Optics, 3rd ed.* (Academic Press, New York, 2001)

[89]   H. A. Haus, "Noise figure definition valid from RF to optical frequencies," IEEE J. Select. Topics Quantum Electron. **6**, 240-247 (2000)

[90]   D. M. Baney, P. Gallion, and R. S. Tucker, "Theory and measurement techniques for the noise figure of optical amplifiers," Optical Fiber Technol. **6**, 122-154 (2000)

[91]   E. Desurvire, *Erbium-doped fiber amplifiers: Principles and applications* (Wiley, New York, 1994)

[92]   N. A. Olsson, "Lightwave systems with optical amplifiers," J. Lightwave Technol. **7**, 1071-1082 (1989)

[93]   A. Borghesani, N. Fensom, A. Scott, G. Crow, L. M. Johnston, J. A. King, L. J. Rivers, S. Cole, S. D. Perrin, D. Scrase, G. Bonfrate, A. D. Ellis, I. F. Lealman, G. Crouzel, L. S. H. K. Chun, A. Lupu, E. Mahe, and P. Maigne, "High saturation power (> 16.5 dBm) and low noise figure (< 6 dB) semiconductor optical amplifier for C-band operation," *Opt. Fiber Commun. Conf.* (OFC'03). Techn. Digest (Atlanta, GA, USA, 2003), paper ThO1

[94]   K. Morito and S. Tanaka, "Record high saturation power (+22 dbm) and low noise figure (5.7 dB) polarization-insensitive soa module," IEEE Photon. Technol. Lett. **17**, 1298-1300 (2005)

[95]   G. P. Agrawal and N. A. Olsson, "Self-phase modulation and spectral broadening of optical pulses in semiconductor laser amplifiers," IEEE J. Quantum Electron. **25**, 2297-2306 (1989)

[96]   R. Olshansky, C. B. Su, J. Manning, and W. Powazinik, "Measurement of radiative and nonradiative recombination rates in InGaAsP and AlGaAs light sources," IEEE J. Quantum Electron. **QE-20**, 838-854 (1984)

[97]   J. Leuthold, G. Raybon, Y. Su, R. Essiambre, S. Cabot, J. Jaques, and M. Kauer, "40 Gbit/s transmission and cascaded all-optical wavelength conversion over 1 000 000 km," Electron. Letters **38**, 890- 892 (2002)

[98]   H. Chen, G. Zhu, Q. Wang, J. Jaques, J. Leuthold, A. B. Piccirilli, and N. K. Dutta, "All-optical logic XOR using differential scheme and Mach-Zehnder interferometer," Electronics Letters **38**, 1271-1273 (2002)

[99]   J. P. Sokoloff, P. R. Prucnal, I. Glesk, and M. Kane, "A terahertz optical asymmetric demultiplexer (TOAD)," IEEE Photon. Technol. Lett. **5**, 787-790 (1993)

[100]   A. Bjarklev, *Optical Fiber Amplifiers: Design and System Applications*, (Artech House, Norwood, 1993)

[101]   L. Occhi, *Semiconductor optical amplifiers made of ridge waveguide bulk InGaAsP/InP: Experimental characterization and numerical modeling of gain, phase and noise*, PhD thesis, ETH Zürich (2002)

[102] R. J. Manning, D. A. O. Davies, and J. K. Lucek, "Recovery rates in semiconductor laser amplifiers: Optical and electrical bias dependencies," Electron. Lett. **30**, 1233-1235 (1994)

[103] F. Girardin, G. Guekos, and A. Houbavlis, "Gain recovery of bulk semiconductor optical amplifiers," IEEE Photon. Technol. Lett. **10**, 784-786 (1998)

[104] J. Slovak, C. Bornholdt, U. Busolt, G. Bramann, Ch. Schmidt, H. Ehlers, H. P. Nolting, and B. Sartorius, "Optically clocked ultra long SOAs: A novel technique for high speed 3R signal regeneration," Opt. Fiber Commun. Conf. (OFC'04). Techn. Digest (Los Angeles, CA, USA, 2004), paper WD4

[105] R. Gutiérrez-Castrejón, L. Schares, L. Occhi, and G. Guekos, "Modeling and measurement of longitudinal gain dynamics in saturated semiconductor optical amplifiers of different length," IEEE J. Quantum Electron. **36**, 1476-1484 (2000)

[106] A. Kapoor, E. K. Sharma, W. Freude, and J. Leuthold, "Investigation of the saturation characteristics of InGaAsP-InP bulk SOA," Proc. SPIE 7597 (2010), 75971I (2010)

[107] J. Mørk, M. L. Nielsen, and T. W. Berg, "The dynamics of semiconductor optical amplifiers, modeling and applications," Optics Photon. News, **14**, 43-48 (2003)

[108] A. Mecozzi and J. Mørk, "Saturation induced by picosecond pulses in semiconductor optical amplifiers," J. Opt. Soc. Am. B **14**, 761-770 (1997)

[109] J. Mørk and A. Mecozzi, "Theory of the ultrafast optical response of active semiconductor waveguides," J. Opt. Soc. Am. B. **13**, 1803-1816 (1996)

[110] A. Mecozzi and J. Mørk, "Saturation effects in nondegenerated four-wave mixing between short optical pulses in semiconductor laser amplifiers," IEEE J. Selec. Topics Quantum Electron. **3**, 1190-1207 (1997)

[111] A. V. Uskov, E. P. O'Reilly, M. Laemmlin, N. N. Ledentsov, and D. Bimberg, "On gain saturation in quantum dot semiconductor optical amplifiers," Opt. Commun. **248**, 211-219 (2005)

[112] J. Wang, A. Maitra, C.G. Poulton, W. Freude, and J. Leuthold, "Temporal dynamics of the alpha factor in semiconductor optical amplifiers," J. Lightwave Technol. **25**, 891-900 (2007)

[113] C. Meuer, GaAs-based Quantum-Dot Semiconductor Optical Amplifiers at 1.3 μm for All-Optical Networks, PhD thesis, University Berlin (2011)

[114] D. Bimberg and U. W. Pohl, "Quantum dots: promises and accomplishments," materials today **14**, 388-397 (2011)

[115] T. Vallaitis, Ultrafast Nonlinear Silicon Waveguides and Quantum Dot Semiconductor Optical Amplifiers: Characterization and Applications, PhD thesis, Karlsruhe Institute of Technology (2011)

[116] A. V. Uskov, J. Mørk, B. Tromberg, T. W. Berg, I. Magnusdottir, and E. P. O'Reilly, "On high-speed cross-gain modulation without pattern effects in quantum dot semiconductor optical amplifiers," Optics Commun. **227**, 363-369 (2003)

[117] A. V. Uskov, E. P. O'Reilly, R. J. Manning, R. P. Webb, D. Cotter, M. Laemmlin, N. N. Ledentsov, and D. Bimberg, "On ultrafast switching based on quantum-dot semiconductor optical amplifiers in nonlinear interferometers," IEEE Photon. Technol. Lett. **16**, 1265-1267 (2004)

[118] J. Wang, A. Marculescu, J. Li, P. Vorreau, S. Tzadok, S. Ben Ezra, S. Tsadka, W. Freude, and J. Leuthold, "Pattern effect removal technique for semiconductor-optical-amplifier-based wavelength conversion," IEEE Photon. Technol. Lett. **19**, 1955-1957 (2007)

[119] A. V. Uskov, T. W. Berg, and J. Mørk, "Theory of Pulse-Train Amplification Without Patterning Effects in Quantum-Dot Semiconductor Optical Amplifiers," IEEE J. Quantum Electr. **40**, 306-320 (2004)

[120] A. Bilenca and G. Eisenstein, "On the noise properties of linear and nonlinear quantum-dot semiconductor optical amplifiers: The impact of inhomogeneously broadened gain and fast carrier dynamics," IEEE J. Quantum Electr. **40**, 690-702 (2004)

[121] T. W. Berg, J. Mork, J. M. Hvam, "Gain dynamics and saturation in semiconductor quantum dot amplifiers," New. J. Phys. 6, 178-184 (2004)

[122] C. Koos, *Nanophotonic Devices for Linear and Nonlinear Optical Signal Processing*, PhD thesis, University of Karlsruhe (TH) (2007)

[123] C. Dorrer and I. Kang, "Real-time implementation of linear spectrograms for the characterization of high bit rate optical pulse trains," Photon. Techn. Lett. **16**, 858-860 (2004)

[124] D. Wolfson, S. L. Danielsen, C. Joergensen, B. Mikkelsen, and K. E. Stubkjaer, "Detailed theoretical investigation of the input power dynamic range for gain-clamped semiconductor optical amplifier gates at 10 Gb/s," IEEE Photon. Technol. Lett. **10**, 1241-1243 (1998)

[125] D. A. Francis, S. P. DiJaili, and J. D. Walker, "A single-chip linear optical amplifier," in *Proc. Optical Fiber Communication Conference* (OFC'01), Anaheim (CA), USA, 17.-22.03.2001 (2001), postdeadline paper PDP 13-1

[126] C. Michie, A. E. Kelly, I. Armstrong, I. Andonovic, and C. Tombling, "An adjustable gain-clamped semiconductor optical amplifier (AGC-SOA)," J. Lightwave Technol. **25**, 1466-1473 (2007)

[127] H. N. Tan, M. Matsuura, and N. Kishi, "Enhancement of input power dynamic range for multiwavelength amplification and optical signal processing in a semiconductor optical amplifier using holding beam effect," J. Lightwave Technol. **8**, 2593-2602 (2010)

[128] J. Yu and P. Jeppesen, "Increasing input power dynamic range of SOA by shifting the transparent wavelength of tunable optical filter," J. Lightwave Technol. **19**, 1316-1325 (2001)

[129] J. Leuthold, D. Marom, S. Cabot, R. Ryf, P. Bernasconi, F. Baumann, J. Jaques, D. T. Neilson, and C. R. Gile, "All-optical wavelength converter based on a pulse reformatting optical filter", *Proc. Optical Fiber Communications Conference* (OFC'2003), Atlanta, USA, Paper PD41, March 2003

[130] A. Fiore and A. Markus, "Differential gain and gain compression in quantum-dot lasers," IEEE J. Quantum Electron. **43**, 287-294 (2007)

[131] P. Runge, *Nonlinear Effects in Ultralong Semiconductor Optical Amplifiers for Optical Communications: Physics and Applications*, PhD thesis, University of Berlin (TH) (2010)

[132] G. Contestabile, A. Maruta, S. Sekiguchi, K. Morito, M. Sugawara, and K. Kitayama, "Cross-Gain Modulation in Quantum-Dot SOA at 1550 nm," IEEE J. Quantum Electron. **46**, 1696-1703 (2010)

[133] G.-W. Lu, M. Sköld, P. Johannisson, J. Zhao, M. Sjödin, H. Sunnerud, M. Westlund, A. Ellis, and P. A. Andrekson, "40-Gbaud 16-QAM transmitter using tandem IQ modulators with binary driving electronic signals," Opt. Express **18**, 23062-23069 (2010)

[134] R. Giller, R. J. Manning, and D. Cotter, "Gain and phase recovery of optically excited semiconductor optical amplifiers," IEEE Photon. Technol. Lett **18**, 1061-1063 (2006)

[135] L. G. Kazovsky, W.-T. Shaw, D. Gutierrez, N. Cheng, and S.-W. Wong, "Next-Generation Optical Access Networks," J. Lightwave Technol. **25**, 3428-3442 (2007)

[136] E. Trojer, S. Dahlfort, D. Hood, and H. Mickelsson, "Current and next-generation PONs: A technical overview of present and future PON technology," Ericsson Rev. **2**, 64-69 (2008)

[137] R. Davey, J. Kani, F. Bourgart, and K. McCammon, "Options for Future Optical Access Networks," IEEE Commun. Mag. **44**, 50-56 (2006)

[138] K.-I. Suzuki, Y. Fukada, D. Nesset, and R. Davey, "Amplified gigabit PON systems," J. Opt. Networking **6**, 422-432 (2007)

[139] L. H. Spiekman, J. M. Wiesenfeld, A. H. Gnauck, L. D. Garrett, G. N. van den Hoven, T. van Dongen, M. J. H. Sander-Jochem, and J. J. M. Binsma, "8 x 10 Gb/s DWDM transmission over 240 km of standard fiber using a cascade of semiconductor optical amlifiers," IEEE Photonics Technol. Lett. **12**, 1082-1084 (2000)

[140] P. P. Iannone, K. C. Reichmann, C. Brinton, J. Nakagawa, T. Cusick, E.M. Kimber, C. Doerr, L. L. Buhl, M. Capuzzo, E. Y. Chen, L. Gomez, J. Johnson, A. M. Kanan, J. Lentz, Y. F. Chang, B. Pálsdóttir, T. Tokle, and L. Spiekman, "Bi-Directionally Amplifier Extended Reach 40 Gb/s CWDM-TDM PON with Burst-Mode Upstream Transmission," in *Proc. Optical Fiber Communication Conference* (OFC'11), Los Angeles (CA), USA, 06.-10.03.2011 (2011), paper PDPD6

[141] D. Bimberg, "Quantum dot based nanophotonics and nanoelectronics," Electron. Lett. **44**, 168-170 (2008)

[142] S. Norimatsu, and M. Maruoka, "Accurate Q-Factor Estimation of Optically Amplified Systems in the Presence of Waveform Distortions," J. Lightwave Technol. **20**, 19-27 (2002)

[143] X. Wei, and L. Zhang, "Analysis of the Phase Noise in Saturated SOAs for DPSK Applications," IEEE J. Quantum Electron. **41**, 554-561 (2005)

[144] A. E. Kelly, C. Michie, I. Armstrong, I. Andonovic, C. Tombling, J. McGenough, and B. C. Thomsen, "High-performance semiconductor optical amplifier modules at 1300nm," IEEE Photonics Technol. Lett. **18**, 2674-2676 (2006)

[145] S. Tanaka, S.-H. Jeong, S. Yamazaki, A. Uetake, S. Tomabechi, M. Ekawa, and K. Morito, "Monolithically integrated 8:1 SOA gate switch with large extinction ratio and wide input power dynamic range," IEEE J. Quantum Electron. **45**, 1155-1162 (2009)

[146] R. J. Manning, A. D. Ellis, A. J. Poustie, and K. J. Blow, "Semiconductor laser amplifiers for ultrafast all-optical signal processing," J. Opt. Soc. Am. B **14**, 3204-3216 (1997)

[147] C. Meuer, J. Kim, M. Laemmlin, S. Liebich, A. Capua, G. Eisenstein, A. R. Kovsh, S. S. Mikhrin, I. L. Krestnikov and D. Bimberg, "Static gain saturation in quantum dot semiconductor optical amplifiers," Opt. Express **16**, 8269-8279 (2008)

[148] R. Alizon, D. Hadass, V. Mikhelashvili, G. Eisenstein, R. Schwertberger, A. Somers, J. P. Reithmaier, A. Forchel, M. Calligaro, S. Bansropun, and M. Krakowski, "Cross-saturation dynamics in InAs/InP quantum dash optical amplifiers operating at 1550 nm," Electron. Lett. **41**, 266-268 (2005)

[149] M. Sauer and J. Hurley, "Experimental 43 Gb/s NRZ and DPSK performance comparison for systems with up to 8 concatenated SOAs," in *Conference on Lasers and Electro-Optics and Quantum Electronics and Laser Science Conference*, CLEO/QELS 2006, (2006), paper CThY2

[150] E. Ciaramella, A. D'Errico, and V. Donzella, "Using semiconductor-optical amplifiers with constant envelope WDM signals," IEEE J. Quantum Electron. **44**, 403-409 (2008)

[151] J. D. Downie and J. Hurley, "Effects of dispersion on SOA nonlinear impairments with DPSK signals," in *Proc. of 21st Annual Meeting of the IEEE Lasers and Electro-Optics Society*, LEOS 2008, paper WX3

[152] P. S. Cho, Y. Achiam, G. Levy-Yurista, M. Margalit, Y. Gross, and J. B. Khurgin, "Investigation of SOA nonlinearities on the amplification of high spectral efficiency signals," in *Optical Fiber Communication Conference*, OSA Technical Digest (CD) (Optical Society of America, 2004), paper MF70

[153] X. Wei, Y. Su, X. Liu, J. Leuthold, S. Chandrasekhar, "10-Gb/s RZ-DPSK transmitter using a saturated SOA as a power booster and limiting amplifier," IEEE Photon. Technol. Lett. **16**, 1582-1584 (1998)

[154] H. Takeda, N. Hashimoto, T. Akashi, H. Narusawa, K. Matsui, K. Mori, S. Tanaka, and K. Morito, "Wide range over 20 dB output power control using semiconductor optical amplifier for 43.1 Gbps RZ-DQPSK signal," *35th European Conference on Optical Communication*, 2009. ECOC 2009, paper 5.3.4

[155] V. J. Urick, J. X. Qiu, and F. Bucholtz, "Wide-Band QAM-Over-Fiber Using Phase Modulation and Interferometric Demodulation," IEEE Photon. Technol. Lett. **16**, 2374-2376 (2004)

[156] R. Schmogrow, B. Nebendahl, M. Winter, A. Josten, D. Hillerkuss, S. Koenig, J. Meyer, M. Dreschmann, M. Huebner, C. Koos, J. Becker, W. Freude, and J. Leuthold, "Error vector magnitude as a performance measure for advanced modulation formats," IEEE Photon. Technol. Lett. **24**, 61-63 (2012)

[157] R. Schmogrow, D. Hillerkuss, M. Dreschmann, M. Huebner, M. Winter, J. Meyer, B. Nebendahl, C. Koos, J. Becker, W. Freude, and J. Leuthold, "Real-time software-defined multiformat transmitter generating 64 QAM at 28 GBd," IEEE Photon. Technol. Lett. **22**, 1601-1603 (2010)

[158] F. Ginovart, J. C. Simon, and I. Valiente, "Gain recovery dynamics in semiconductor optical amplifier," Opt. Commun. **199**, 111-115 (2001)

[159] A. A. M. Saleh and I. M. I. Habbab, "Effects of semiconductor-optical-amplifier nonlinearity on the performance of high-speed intensity-modulation lightwave systems," IEEE Trans. Commun. **38**, 839-846 (1990)

[160] K.-P. Ho, "The effect of interferometer phase error on direct-detection DPSK and DQPSK signals," IEEE Photon. Technol. Lett. **16**, 308-310 (2004)

[161] H. Kim, and P. J. Winzer, "Robustness to laser frequency offset in direct-detection DPSK and DQPSK systems," J. Lightwave Technol. **21**, 1887-1891 (2003)

[162] N. J. Gomes, M. Morant, A. Alphones, B. Cabon, J. E. Mitchell, C. Lethien, M. Csörnyei, A. Stöhr, and S. Iezekiel, "Radio-over-fiber transport for the support of wireless broadband services," J. Opt. Netw. **8**, 156-178 (2009)

[163] R. Boula-Picard, M. Alouini, J. Lopez, N. Vodjdani, and J.-C. Simon, "Impact of the Gain Saturation Dynamics in Semiconductor Optical Amplifiers on the Characteristics of an Analog Optical Link," J. Lightwave Technol. **23**, 2420-2426 (2005)

[164] Th. Pfeiffer, "Converged Heterogeneous Optical Metro-Access Networks," *Europ. Conf. Opt. Commun.* (ECOC'10) (Torino, Italy, 2010), paper Tu.5.B.1

[165] N. Antoniades, K. C. Reichmann, P. P. Iannone, and A. M. Levine, "Engineering Methodology for the Use of SOAs and CWDM Transmission in the Metro Network Environment," in *Opt. Fiber Commun. Conf. and Nat. Fiber Opt. Eng. Conf.* (OFC/NFOEC'06), Techn. Digest, paper OTuG6, 2006

[166] S. Liu, K. A. Williams, T. Lin, M. G. Thompson, C. K. Yow, A. Wonfor, R. V. Penty, I. H. White, F. Hopfer, M. Lämmlin, and D. Bimberg, "Cascaded Performance of Quantum Dot Semiconductor Optical Amplifier in a Recirculating Loop," in *Conference on Lasers and Electro-Optics/Quantum Electronics and Laser Science Conference and Photonic Applications Systems Technologies*, Technical Digest (CD) (Optical Society of America, 2006), paper CTuM4

[167] L. Provost, F. Parmigiani, P. Petropoulos, and D. J. Richardson, "Investigation of Simultaneous 2R Regeneration of Two 40-Gb/s Channels in a Single Optical Fiber," Photon. Technol. Lett. **20**, 270-272 (2008)

[168] S. J. B. Yoo, "Wavelength conversion technologies for WDM network applications," IEEE J. Lightwave Technol. **14**, 955-966 (1996)

[169] A. M. Saleh and J. M. Simmons, "Evolution towards the next-generation core optical network," J. Lightwave Technol. **24**, 3303-3321 (2006)

[170] S. Sygletos, I. Tomkos, and J. Leuthold, "Technology challenges on the road towards transparent networking," J. Opt. Netw. **7**, 321-350 (2008)

[171] Y. Ueno, S. Nakamura, K. Tarima, and S. Kitamura, "3.8-THz wavelength conversion of picosecond pulses using a semiconductor delayed-interference signal-wavelength converter (DISC)," IEEE Photon. Technol. Lett. **10**, 346-348 (1998)

[172] Y. Liu, E. Tangdiongga, Z. Li, H. de Waardt, A. M. J. Koonen, G. D. Khoe, Shu Xuewen Shu, I. Bennion, and H. J. S. Dorren, "Error-Free 320-Gb/s All-Optical Wavelength Conversion Using a Single Semiconductor Optical Amplifier," IEEE J. Lightwave Technol. **25**, 103-108 (2007)

[173] M. L. Nielsen, B. Lavigne, and B. Dagens, "Polarity-preserving SOA-based wavelength conversion at 40 Gb/s using bandpass filtering," Electron. Lett. **39**, 1334-1335 (2003)

[174] Y. Liu, E. Tangdiongga, Z. Li, S. Zhang, H. de Waardt, G. D. Khoe, and H. J. S. Dorren, "80 Gbit/s wavelength conversion using semiconductor optical amplifier and optical bandpass filter," Electron. Lett. **41**, 487-489 (2005)

[175] J. Leuthold, B. Mikkelsen, G. Raybon, C. H. Joyner, J. L. Pleumeekers, B. I. Miller, K. Dreyer, and R. Behringer, "All-Optical Wavelength Conversion Between 10 and 100 Gb/s with SOA Delayed-Interference Configuration", Opt. and Quantum Electron. **33**, 939-952 (2001)

[176] S. Nakamura, Y. Ueno, and K. Tajima, "168-Gb/s all-optical wavelength conversion with a symmetric-Mach-Zehnder-type switch," IEEE Photon. Technol. Lett. **13**, 1091-1093 (2001)

[177] P. Borri, W. Langbein, J. M. Hvam, F. Heinrichsdorf, M.-H. Mao, and D. Bimberg, "Ultrafast gain dynamics in InAs-InGaAs quantum dot amplifiers," IEEE Photon. Technol. Lett. **12**, 594-596 (2000)

[178] M. Sugawara, T. Akiyama, N. Hatori, Y. Nakata, H. Ebe, and H. Ishikawa, "Quantum-dot semiconductor optical amplifiers for high-bit rate signal processing up to 160 Gb/s and a new scheme of 3R regenerators," Meas. Sci. Technol. **13**, 1683-1691 (2002)

[179] A. Bilenca, R. Alizon, V. Mikhelashhvili, D. Dahan, G. Eisenstein, R. Schwertberger, D. Gold, J. P. Reithmaier, A. Forchel, "Broad-band wavelength conversion based on cross gain modulation and four wave mixing in InAs-InP quantum dash semiconductor optical amplifiers operating at 1550 nm," IEEE Photon. Technol. Lett. **15**, 563-565 (2003)

[180] T. Akiyama, H. Kuwatsuka, N. Hatori, Y. Nakata, H. Ebe, and M. Sugawara, "Symmetric highly efficient (~0 dB) wavelength conversion based on four wave mixing in quantum dot optical amplifiers," IEEE Photon. Technol. Lett. **14**, 1139-1141 (2002)

[181] O. Raz, J. Herrera, N. Calabretta, E. Tangdiongga, A. Anantathanasarn, R. Nötzel, and H. J. S. Dorren, "Non-inverted multiple wavelength converter at 40 Gbit/s using 1550 nm quantum dot SOA," Electron Lett. **44**, 988-989 (2008)

[182] G. Contestabile, A. Maruta, S. Sekiguchi, K. Morito, M. Sugawara, and K. Kitayama, "160 Gb/s cross gain modulation in quantum dot SOA at 1550 nm," _35th European Conf. Opt. Commun._ (ECOC'09), Vienna, Austria, PDP, 2009

[183] M. L. Nielsen and J. Mørk, "Experimental and theoretical investigation of the impact of ultra-fast carrier dynamics on high speed SOA-based all-optical switches," Opt. Express **14**, 331-347 (2005)

[184] T. von Lerber, S. Honkanen, A. Tervonen, H. Ludvigsen, and F. Kueppers, "Optical clock recovery methods: Review (Invited)," Opt. Fiber Technol. **15**, 363-372 (2009)

[185] C. Ware, L. K. Oxenløwe, F. G. Agis, H. C. Mulvad, M. Galili, S. Kurimura, H. Nakajima, I. Ichikawa, D. Erasme, A. T. Clausen, and P. Jeppesen, "320 Gbps to 10 GHz sub-clock receevery using a PPLN-based opto-electronic phase-locked loop," Opt. Express **16**, 5007-5012 (2008)

[186] L. K. Oxenløwe, F. Agis, C. Ware, S. Kurimura, H. Mulvad, M. Galili, K. Kitamura, H. Nakajima, J. Ichikawa, D. Erasme, A. Clausen, and P. Jeppesen, "640 Gbit/s clock recovery using periodically poled lithium niobate," Electron. Lett. **44**, 370-371 (2008)

[187] M. Jinno, T. Matsumoto, and M. Koga, "All-optical timing extraction using an optical tank circuit," IEEE Photon. Technol. Lett. **2**, 203-204 (1990)

[188] V. Roncin, S. Lobo, L. Bramerie, A. O'Hare, and J.-C. Simon, "System characterization of a passive 40 Gb/s all optical clock recovery ahead of the receiver," Opt. Express **15**, 6003-6009 (2007)

[189] T. Wang, C. Lou, L. Huo, Z. Wang, and Y. Gao, "A simple method for clock recovery," Opt. Laser Technol. **36**, 613-616 (2004)

[190] G. Contestabile, A. D'Errico, M. Presi, and E. Ciaramella, "40-Ghz all-optical clock extraction using a semiconductor-assisted Fabry–Pérot filter," IEEE Photon. Technol. Lett. **16**, 2523-2525 (2004)

[191] G. Contestabile, M. Presi, N. Calabretta, and E. Ciaramella, "All-optical clock recovery for NRZ-DPSK signals," IEEE Photon. Technol. Lett. **18**, 2544-2546 (2006)

[192] M. Presi, N. Calabretta, G. Contestabile, and E. Ciaramella, "Wide dynamic range all-optical clock and data recovery from preamble-free NRZ-DPSK packets," IEEE Photon. Technol. Lett. **19**, 372-374 (2007)

[193] J. Lee, H. Cho, S. K. Lim, S. S. Lee, and J. S. Ko, "Optical clock recovery scheme for a high bit rate nonreturn-to-zero signal using fiber Bragg grating filters," Opt. Eng. **44**, 020502 (2005)

[194] J. Lee, H. Cho, and J. S. Ko, "Enhancement of clock component in a nonreturn-to-zero signal through beating process," Opt. Fiber Technol. **12**, 59–70 (2006)

[195] T. Akiyama, H. Kuwatsuka, T. Simoyama, Y. Nakata, K. Mukai, M. Sugawara, O. Wada, and H. Ishikawa, "Nonlinear gain dynamics in quantum-dot optical amplifiers and its application to optical communication devices," EEE J. Quantum Electron. **37**, 1059-1065 (2001)

[196] C. Johnson, K. Demarest, C. Allen, R. Hui, K. V. Peddanarappagari, and B. Zhu, "Multiwavelength all-optical clock recovery," IEEE Photon. Technol. Lett. **11**, 895-897 (1999)

[197] D. Pudo, M. Depa, and L. Chen, "Single and multiwavelength all-optical clock recovery in single-mode fiber using the temporal Talbot effect," J. Lightwave Technol. **25**, 2898-2903 (2007)

[198] V. Mikhailov and B. Payvel, "All-optical multiwavelength clock recovery using integrated semiconductor amplifier module," Electron. Lett. **37**, 232-234 (2001)

[199] F. Wang, Y. Y. Huang and X. Zhang, "Single and multiwavelength all-optical clock recovery using Fabry-Pérot Semiconductor Optical Amplifier," IEEE Photon. Technol. Lett. **21**, 1109-1111 (2009)

[200] M. Spyropoulou, N. Pleros, G. Papadimitriou, and I. Tomkos, "A high speed multi-wavelength clock recovery scheme for optical packets," IEEE Photon. Technol. Lett. **20**, 2147-2149 (2008)

[201] C. R. Doerr, S. Chandrasekhar, P. J. Winzer, L. Stultz, A. R. Chraplyvy, and R. Pafcheck, "Simple multi-channel optical equalizer for mitigating intersymbol interference," *in Proc. OFC'03*, Atlanta, USA Paper PD 11

[202] P. Vorreau, A. Marulescu, J. Wang, G. Böttcher, B. Sartorius, C. Bornholdt, J. Slovak, M. Schlak, C. Schmidt, S. Tsadka, W. Freude, and J. Leuthold, "Cascadability and regenerative properties of SOA all-optical DPSK wavelength converters," IEEE Photon. Technol. Lett. **18**, 1970-1972 (2006)

[203] R. P. Webb, X. Yang, R. J. Manning, G. D. Maxwell, A. J. Poustie, S. Lardenois, and D. Cotter,"All-optical binary pattern recognition at 42 Gb/s," IEEE J. Lightwave Technol. **27**, 2240-2245 (2009)

[204] P. J. Winzer, A. H. Gnauck, G. Raybon, S. Chandrasekhar, Y. Su, and J. Leuthold, "40-Gb/s return-to-zero alternate-mark-inversion (RZ-AMI) transmission over 2000 km," IEEE Photon. Technol. Lett. **15**, 766-768 (2003)

[205] K. Inoue, "Noise transfer function characteristics in wavelength conversion based on cross-gain saturation in a semiconductor optical amplifier," IEEE Photon. Technol. Letters. **8**, 888-890 (1996)

[206] S. I. Pegg, M. J. Fice, M. Adams, and A. Hadjifotiou, "Noise in wavelength conversion by cross gain modulation in a semiconductor optical amplifier," IEEE Photon. Technol. Letters. **11**, 724-726 (1999)

[207] T. Akiyama, N. Hatori, Y. Nakata, H. Ebe, and M. Sugawara, "Pattern-effect-free amplification and cross-gain modulation achieved by using ultrafast gain nonlinearity in quantum-dot semiconductor optical amplifiers," Physica Status Solidi B **238**, 301-304 (2003)

[208] G. Contestabile, A. Maruta, S. Sekiguchi, K. Morito, and K. Kitayama, "80 Gb/s Multicast Wavelength Conversion by XGM in a QD-SOA," in *European Conference on Optical Communication* (ECOC2010)(Torino, Italy, 2010), paper Mo.2.A.3

[209] J. Kim, C. Meuer, D. Bimberg, and G. Eisenstein, "Role of carrier reservoirs on the slow phase recovery of quantum dot semiconductor optical amplifiers," Appl. Phys. Lett. **94**, 41112-41114 (2009)

[210] I. Kang, C. Dorrer, L. Zhang, M. Dinu, M. Rasras, L. L. Buhl, S. Cabot, A. Bhardwaj, X. Liu, M. A. Cappuzzo, L. Gomez, A. Wong-Foy, Y. F. Chen, N. K. Dutta, S. S. Patel, D. T. Neilson, C. R. Giles, A. Piccirilli, and J. Jaques, "Characterization of the dynamical processes in all-optical signal processing using semiconductor optical amplifiers," IEEE J. Sel. Top. in Quantum Electron. **14**, 758-769 (2008)

[211] D. F. Geraghty, R. B. Lee, M. Verdiell, M. Ziari, A. Mathur, and K. J. Vahala, "Wavelength conversion for WDM communication systems using four-wave mixing in semiconductor optical amplifiers," IEEE J. Sel. Top. in Quantum Electron. **3**, 1146-1155 (1997)

[212] C. Meuer, C. Schmidt-Langhorst, H. Schmeckebier, G. Fiol, D. Arsenijević, C. Schubert, and D. Bimberg, "40 Gb/s wavelength conversion via four-wave mixing in a quantum-dot semiconductor optical amplifier," Opt. Express **19**, 3788-3798 (2011)

[213] N. Ghani, J.-Y. Pan, and X. Cheng, "Metropolitan Optical Networks, Optical Fiber Telecommunications, vol. IVB", Academic Press 2002, pp. 329-403

[214] A. Gladisch, R.-P Braun, D. Breuer, A. Ehrhardt, H.-M. Foisel, M. Jaeger, R. Leppla, M. Scheiders, S. Vorbeck, W. Weiershausen, and F.-J. Westphal, "Evolution of Terrestrial Optical System and Core Network Architecture," Proc. IEEE **94**, 869-891 (2006)

[215] A.Morea and J. Poirrier, "A Critical Analysis of the Possible Cost Savings of Translucent Networks," in *Proceedings of 5th international Workshop on Design of Reliable Communication Networks*, 311-317 (2005)

[216] P. Humblet, "The direction of optical technology in the metro area," presented at *OSA Conf. Opt. Fiber Commun. 2001* (OFC' 01), March 2001, paper WBB1

[217] J. Livas, "Optical Transmission Evolution: From Digital to Analog to ? Network Tradeoffs Between Optical Transparency and Reduced Regeneration Cost," J. Lightwave Technol. **23**, 219-224 (2005)

[218] O. Leclerc, "Towards Transparent Optical Networks: still some challenges ahead," in *Proceedings 18th Annual Meeting of the IEEE Lasers and Electro-Optics Society*, paper TuCC3 (invited), 418-419, 2005

[219] R. Nejabati, G. Zervas, G. Zarris, Y. Qin, E. Escalona, M. O'Mahony, and D. Simeonidou," Multigranular optical router for future networks [invited]," J. Opt. Netw. **7**, 914-927, (2008)

[220] R. S. Barr and R. A. Patterson, "Grooming Telecommunications Networks," Optical Networks Magazine **2**, 20-23 (2001)

[221] B. Mukherjee, C. Ou, H. Zhu, N. Singhal, and S. Yao, "Traffic Grooming in Mesh Optical Networks," *presented at OSA Conf. Opt. Fiber Commun. 2004* (OFC'04), Feb. 2004, ThG1

[222] S. K. Ibrahim, R. Weerasuriya, D. Hillerkuss, G. Zarris, D. Simeonidou, J. Leuthold, D. Cotter, and A. Ellis, "Experimental demonstration of 42.6Gbit/s asynchronous digital optical regenerators," *presented at Int. Conf. Transparent Opt. Netw. 2008* (ICTON '08), June 2008, We.C3.3

[223] D. Cotter and A. D. Ellis, "Asynchronous Digital Optical Regeneration and Networks," J. Lightwave Technol. **16**, 2068-2080, (1998)

[224] D. Hillerkuss, A. Ellis, G. Zarris, D. Simeonidou, J. Leuthold, and D. Cotter, "40Gbit/s asynchronous digital optical regenerator," Opt. Express **16**, 18889-18894 (2008)

[225] G. Zarris, P. Vorreau, D. Hillerkuss, S. K. Ibrahim, R. Weerasuriya, A. D. Ellis, J. Leuthold, and D. Simeonidou, "WDM-to-OTDM Traffic Grooming by means of Asynchronous Retiming," *presented at OSA Conf. Opt. Fiber Commun. 2009* (OFC'09), March 2009, OThJ6

[226] R. Morais, R. Meleiro, P. Monteiro, and P. Marques, "OTDM-to-WDM Conversion based on Wavelength Conversion and Time Gating in a Single Optical Gate," *presented at OSA Conf. Opt. Fiber Commun. 2007 (OFC'07)*, March 2007, OTuD5.

[227] P. V. Mamyshev, "All-optical data regeneration based on self-phase modulation effect," *presented at Eur. Conf. Opt. Commun. 1998* (ECOC'98), Sep. 1998, 475-476.

[228] S. K. Ibrahim, D. Hillerkuss, R. Weerasuriya, G. Zarris, D. Simeonidou., J. Leuthold, and A. D. Ellis, "Novel 42.65 Gbit/s dual gate asynchronous digital optical regenerator using a single MZM," *presented at Eur. Conf. Opt. Commun. 2008* (ECOC'08), Sept. 2008, Tu4D3.

[229] G. Zarris, E. Hugues-Salas, N. Amaya Gonzalez, R. Weerasuriya, ,F. Parmigiani, D. Hillerkuss, P. Vorreau, M. Spyropoulou, S. K. Ibrahim, A. D. Ellis, R. Morais, P. Monteiro, P. Petropoulos, D. Richardson, I. Tomkos, J. Leuthold, and D. Simeonidou, "Field experiments with a grooming switch for OTDM meshed networking," J. Lightwave Technol. **28**, 316-327 (2010)

[230] C. K. Madsen and J. H. Zhao, *Optical Filter Design and Analysis* (John Wiley & Sons, New York, 1999)

[231] J. Li, K. Worms, R. Maestle, D. Hillerkuss, W. Freude, J. and Leuthold, "Free-space optical delay interferometer with tunable delay and phase," Opt. Express **19**, 11654-11666 (2011)

[232] W. Freude, *Optical Transmitters and Receivers* (Institute of Photonics and Quantum Electronics, Karlsruhe, Germany, 2011), Lecture notes

[233] J. Armstrong, "OFDM for Optical Communications," J. Lightwave Technol. **27**, 189-204 (2009)

[234] "Corning SMF-28 Optical Fiber," Corning, Inc., New York, NY, USA, product information (2002) [online]

[235] M. P. Dlubek, A. J. Phillips, and E. C. Larkins, "Optical signal quality metric based on statistical moments and Laguerre expansion," Qpt. Quant. Electron. **40**, 561-575 (2008)

[236] O. K. Tonguz and L. G. Kazovsky, "Theory of Direct-Detection Lightwave Receivers using Optical Amplifiers," J. Lightwave Technol. **9**, 174-181 (1991)

[237] N. S. Bergano, F. W. Kerfoot, and C. R. Davidson, "Margin Measurements in Optical Amplifier Systems," IEEE Photon. Technol. Lett. **5**, 304-306 (1993)

[238] F. Öhman, *Optical Regeneration and Noise in Semiconductor Devices*, PhD thesis, Research Center COM, Technical University of Denmark (2005)

[239] X. Zhang, G. Zhang, C. Xie, and L. Wang, "Noise statistics in optically preamplified differential phase-shift keying receivers with Mach-Zehnder interferometer demodulation," Opt. Lett. **29**, 337-339 (2004)

[240] K.-P Ho, *Phase-modulated Optical Communication Systems* (Springer, 2005)

[241] W. Hong, D. Huang, X. Zhang, and G. Zhu, "Simulation and evaluation of phase noise for optical amplification using semiconductor optical amplifiers in DPSK applications," Opt. Commun. **281**, 28-36 (2008)

[242] C. Dorrer and I. Kang, "Simultaneous temporal characterization of telecommunication optical pulses and modulators by use of spectrograms," Opt. Lett. **27**, 1315-1317 (2002)

Acknowledgements (German)

Die vorliegende Dissertation entstand während meiner Tätigkeit am Institut für Photonik und Quantenelektronik (IPQ) des Karlsruher Instituts für Technologie (KIT). Sie war zum Teil eingebunden in die Europäischen Forschungsprojekte TRIUMPH und Euro-Fos, das vom Bundesministerium für Bildung und Forschung geförderten BMBF-Projektes CONDOR sowie in die Karlsruhe School of Optics and Photonics (KSOP). Am Ende meiner Arbeit möchte ich all denjenigen Personen meinen herzlichen Dank aussprechen, die im Verlauf der letzten Jahre zum Gelingen des vorliegenden Manuskriptes beigetragen haben.

An erster Stelle danke ich meinem Doktorvater Prof. Dr. sc. nat. J. Leuthold für das entgegengebrachte Vertrauen, das große und stetige Interesse an meiner Arbeit, die zahlreichen Anregungen und seine innovativen Ideen. Sein fachlicher Überblick, die zahlreichen Diskussionen über die optischen Halbleiterverstärker und das TRIUMPH Projekt sowie sein Blick für das große Ganze trugen essentiell zum Gelingen der Arbeit bei.

Herrn Prof. Dr. Dr. h.c. W. Freude, meinem zweiten Betreuer, danke ich für die akribische Durchsicht meiner vielen Manuskripte. Seine vorbildliche Genauigkeit und der Drang zur Perfektion ist eine unvergessliche Schule fürs weitere Leben. Seine große fachliche Kompetenz und seine Erfahrung gewährleistete immer ein hohes wissenschaftliches Niveau.

Prof. Dr. rer. nat. Uli Lemmer danke ich für die freundliche Übernahme des Korreferats.

Meinen Kollegen Dr. Thomas Vallaitis und Swen König danke ich für die langjährige erfolgreiche Zusammenarbeit. Die vielen gemeinsamen Messungen und unzähligen Diskussionen über physikalische wie elektrotechnische Zusammenhänge bildeten ein sehr wichtiges Fundament zur Durchführung und dem Gelingen der Arbeit. Das gegenseitige uneingeschränkte Vertrauen im Umgang mit Ideen und Veröffentlichungen ließen eine Teamarbeit auch unter höchster Belastung zur Freude werden.

Dr. Stelios Sygletos danke ich für die Unterstützung im Bereich der Modellierung der optischen Quantenpunktverstärker und für die unzähligen Diskussionen über Simulationsergebnisse und der dazugehörigen messtechnischen Realität.

Die vorliegende Arbeit wurde auch durch sehr fruchtbare Kooperationen begleitet. Ich danke Prof. Dr. D. Bimberg (TU Berlin), Dr. G. Duan, Dr. R. Brenot und Dr. F. Lelarge (alle III-V Labs Alcatel-Lucent) für die Überlassung von optischen Quantenpunkthalbleiterverstärkern und die zahlreichen Diskussionen.

Dr. Christian Meuer (TU Berlin, HHI Berlin) danke ich für die harten und sehr konzentrierten Nachtschichten während unserer gemeinsamen Messungen. Unvergessen werden mir „Himbeereis zum Frühstück", der nächtliche Fleischkäse und die gemeinsamen „Kais Pizza-Schlachten" im Gedächtnis bleiben. Der „MeuRer" war und ist mir immer ein sehr angenehmer und kompetenter Ansprechpartner, der mit vielen Ideen und praktischen Lösungsvorschlägen jede Messung voranbrachte.

Holger Schmeckebier (TU Berlin), genannt „Lefty", danke ich für die gemeinsamen Messungen, die vielen Diskussionen und die akribische Arbeit im Labor.

Dr. Carsten Schmidt-Langhorst und Dr. Colja Schubert vom Heinrich Hertz Institut (HHI) in Berlin sowie Maria Spyropoulou (NTUA) danke ich für die freundliche Aufnahme während der gemeinsamen Messungen im Rahmen des Eurofos-Projektes.

Für die sorgfältige Durchsicht des vorliegenden Manuskriptes bedanke ich mich recht herzlich bei Dr. Christian Meuer, Swen König, Moritz Röger, Dr. Sean O'Duill, Nicole Lindenmann, Alexandra Ludwig, David Hillerkuss, Jing-Shi Li und Manfred Bonk.

Meinen Bürokollegen Dr. Thomas Vallaitis und Argishti Melikyan sowie meinen Bürokolleginnen Nicole Lindenmann, Alexandra Ludwig und Johanna Gütlein danke ich für die freundliche und herzliche Zusammenarbeit über die vielen Jahre, in denen man viel miteinander Durchleben konnte.

Ein großer Dank geht an „meine" zahlreichen motivierten und sehr guten Studenten, die zu dieser Arbeit beigetragen haben: Matthias Hoh, Jörg Pfeifle, Thomas Schellinger, Djorn Karnick, Daniel Lindt, Benedikt Bäuerle, Arne Josten, Akhmet Tussupov, Johanna Gütlein, Gregor Huber und Christian Rode.

Ich möchte auch den Mitarbeitern des IPQs danken, die mich in den vergangenen Jahren mit einer Vielzahl von Arbeiten unterstützt haben. Bei Frau Bernadette Lehmann, Frau Ilse Kober, Frau Angelika Olbrich und Frau Andrea Riemensperger bedanke ich mich für die administrative Unterstützung. Vor allem Bernadette danke ich für die vielen leckeren Kekse und Gummibärchen sowie die Organisation des täglichen Kaffees. Zu Dank bin ich auch Oswald Speck aus dem Packaging-Labor für den sehr kompetenten Umgang mit den Quantenpunktverstärkern und die sehr zuverlässige Organisation von Laborverbrauchsmitteln verpflichtet. Auch Herrn Bürger, Herrn Hirsch und Herrn Höhne aus der mechanischen Werkstatt sage ich ein großes Dankeschön für die vielen Arbeiten, vor allem zu Beginn meiner Arbeit als das IPQ noch aus „leeren" Laboren bestand. Den Herren Martin Winkeler und Sebastian Struck danke ich für die Unterstützung im Bereich Elektronik, Computer und Administration.

Ein herzliches Dankeschön gilt der frühmorgendlichen Fußballrunde des IPQs und IHEs, vor allem meinem Kollegen Moritz Röger für die zuverlässige Organisation.

Bei allen Kollegen des IPQs bedanke ich mich recht herzlich für die gute Zusammenarbeit, Kooperation, den Spaß und die Erfahrungen auch außerhalb des Instituts: Swen König, Moritz Röger, Dr. Thomas Vallaitis, Prof. Dr. C. Koos, Nicole Lindenmann, Alexandra Ludwig, David Hillerkuss, René Schmogrow, Dr. Arvind Mishra, Dr. Sean O'Duill, Dr. Philipp Vorreau, Dr. Stelios Sygletos, Luca Alloatti, Dietmar Korn, Jinshi Li, Christos Klamouris, Argishti Melikyan, Sascha Mühlbrandt, Robert Palmer, Jörg Pfeifle, Philipp Schindler, Simon Schneider, Claudius Weimann, Kai Worms und Frans Wegh.

Ein besonderer Dank gilt Tina, Swen und Moritz sowie Evelyn und Tom für die notwendige Ablenkung neben dem IPQ.

Meinen Eltern Angelika und Manfred-Eckart Bonk sowie meiner Großmutter Margarethe Bachmann danke ich für die uneingeschränkte Unterstützung, auf die ich immer zählen konnte. Desweiteren danke ich der ganzen Familie für die Ermutigungen über all die Jahre.

List of Own Publications

Journal Papers

[J1] **R. Bonk**, G. Huber, T. Vallaitis, S. Koenig, R. Schmogrow, D. Hillerkuss, R. Brenot, F. Lelarge, G.H. Duan, S. Sygletos, C. Koos, W. Freude, and J. Leuthold, "Linear Semiconductor Optical Amplifiers for Amplification of Advanced Modulation Formats," Opt. Express **20**, 9657-9672 (2012)

[J2] **R. Bonk**, T. Vallaitis, J. Guetlein, C. Meuer, H. Schmeckebier, D. Bimberg, C. Koos, W. Freude, and J. Leuthold, "The Input Power Dynamic Range of a Semiconductor Optical Amplifier and Its Relevance for Access Network Applications," IEEE Photonics J. **3**, 1039-1053 (2011)

[J3] D. Hillerkuss, R. Schmogrow, T. Schellinger, M. Jordan, M. Winter, G. Huber, T. Vallaitis, **R. Bonk**, P. Kleinow, F. Frey, M. Roeger, S. Koenig, A. Ludwig, A. Marculescu, J. Li, M. Hoh, M. Dreschmann, J. Meyer, S. Ben Ezra, N. Narkiss, B. Nebendahl, F. Parmigiani, P. Petropoulos, B. Resan, A. Oehler, K. Weingarten, T. Ellermeyer, J. Lutz, M. Moeller, M. Huebner, J. Becker, C. Koos, W. Freude, and J. Leuthold, "26 Tbit s−1 line-rate super-channel transmission utilizing all-optical fast Fourier transform processing, " J. Nature Photon. **5**, 364-371 (2011)

[J4] C. Meuer, C. Schmidt-Langhorst, **R. Bonk**, H. Schmeckebier, D. Arsenijević, G. Fiol, A. Galperin, J. Leuthold, C. Schubert, and D. Bimberg, "80 Gb/s wavelength conversion using a quantum-dot semiconductor optical amplifier and optical filtering," Opt. Express **19**, 5134-5142 (2011)

[J5] **R. Bonk**, P. Vorreau, D. Hillerkuss, W. Freude, G. Zarris, D. Simeonidou, F. Parmigiani, P. Petropoulos, R. Weerasuriya, S. Ibrahim, A. D. Ellis, D. Klonidis, I. Tomkos, and J. Leuthold, "An All-Optical Grooming Switch for Interconnecting Access and Metro Ring Networks," J. Opt. Commun. and Netw. **3**, 206-214 (2011)

[J6] T. Vallaitis, **R. Bonk**, J. Guetlein, D. Hillerkuss, J. Li, R. Brenot, F. Lelarge, G. H. Duan, W. Freude, and J. Leuthold, "Quantum dot SOA input power dynamic range improvement for differential-phase encoded signals," Opt. Express **18**, 6270-6276 (2010)

[J7] S. Sygletos, **R. Bonk**, T. Vallaitis, A. Marculescu, P. Vorreau, J. Li, R. Brenot, F. Lelarge, G. Duan, W. Freude, and J. Leuthold, "Filter assisted wavelength conversion with quantum-dot SOAs," J. Lightwave Technol. **28**, 882-897 (2010)

[J8]   P. Vorreau, S. Sygletos, F. Parmigiani, D. Hillerkuss, **R. Bonk**, P. Petropoulos, D. J. Richardson, G. Zarris, D. Simeonidou, D. Klonidis, I. Tomkos, R. Weerasuriya, S. Ibrahim, A. D. Ellis, D. Cotter, R. Morais, P. Monteiro, S. Ben Ezra, S. Tsadka, W. Freude, and J. Leuthold, "Optical grooming switch with regenerative functionality for transparent interconnection of networks," Opt. Express **17**, 15173-15185 (2009)

[J9]   C. Meuer, J. Kim, M. Laemmlin, S. Liebich, G. Eisenstein, **R. Bonk**, T. Vallaitis, J. Leuthold, A. Kovsh, I. Krestnikov, and D. Bimberg, "High-speed small-signal cross-gain modulation in quantum-dot semiconductor optical amplifiers at 1.3μm," IEEE J. Sel. Top. Quantum Electron. **15**, 749-756 (2009)

[J10] T. Vallaitis, C. Koos, **R. Bonk**, W. Freude, M. Laemmlin, C. Meuer, D. Bimberg, and J. Leuthold, "Slow and fast dynamics of gain and phase in a quantum dot semiconductor optical amplifier," Opt. Express **16**, 170-178 (2008)

[J11] C. Meuer, J. Kim, M. Laemmlin, S. Liebich, D. Bimberg, A. Capua, G. Eisenstein, **R. Bonk**, T. Vallaitis, J. Leuthold, A. R. Kovsh, and I. L. Krestnikov, "40 GHz small-signal cross-gain modulation in 1.3μm quantum dot semiconductor optical amplifiers," Appl. Phys. Lett. **93**, p. 051110 (2008)

[J12] A. D. Ellis, D. Cotter, S. Ibrahim, R. Weerasuriya, C. W. Chow, J. Leuthold, W. Freude, S. Sygletos, P. Vorreau, **R. Bonk**, D. Hillerkuss, I. Tomkos, A. Tzanakaki, C. Kouloumentas, D. J. Richardson, P. Petropoulos, F. Parmigiani, G. Zarris, and D. Simeonidou, "Optical interconnection of core and metro networks," [Invited], J. Opt. Netw. **7**, 928-935 (2008)

Book Publications

[B1]   **R. Bonk**, T. Vallaitis, W. Freude, J. Leuthold, R. V. Penty, A. Borghesani, and I. F. Lealman, "Linear Semiconductor Optical Amplifier," in H. Venghaus, N. Grote (Eds.): "Key Devices in Fiber Optics". Springer Series in Optical Sciences, vol. 161, Berlin, 2012

Conference Contributions

[C1]   J. Leuthold, W. Freude, C. Koos, **R. Bonk**, S. Koenig, D. Hillerkuss, and R. Schmogrow, "Semiconductor Optical Amplifiers in Extended Reach PONs," *R. Slow and Fast Light (SL) Topical Meeting of the OSA*, Toronto, Canada, in Access Networks and In-house Communications, OSA Technical Digest (CD) (Optical Society of America, 2011), paper ATuA1, June 2011 [invited]

[C2]   W. Freude, D. Hillerkuss, T. Schellinger, R. Schmogrow, M. Winter, T. Vallaitis, **R. Bonk**, A. Marculescu, J. Li, M. Dreschmann, J. Meyer, S. Ben Ezra, J. M. Caspi, B. Nebendahl, F. Parmigiani, P. Petropoulos, B. Resan, A. Oehler, K. Weingarten, T. Ellermeyer, J. Lutz, M. Möller, M. Huebner, J. Becker, C. Koos, and J. Leuthold, "All-optical Real-time OFDM Transmitter and Receiver," *J. Conf. on Lasers and Electro-Optics* (CLEO/IQEC'11), Baltimore (MD), Paper CTh01, May 2011

[C3]   N. Lindenmann, I. Kaiser, G. Balthasar, **R. Bonk**, D. Hillerkuss, W. Freude, J. Leuthold, and C. Koos, "Photonic Waveguide Bonds-A Novel Concept for Chip-to-Chip Interconnects," *Proc. Optical Fiber Communication Coference* (OFC'11), Los Angeles (CA), USA, Paper PDPC1, March 2011

[C4]   **R. Bonk**, G. Huber, T. Vallaitis, R. Schmogrow, D. Hillerkuss, C. Koos, W. Freude, and J. Leuthold, "Impact of alfa-factor on SOA Dynamic Range for 20GBd BPSK, QPSK and 16-QAM Signals," *Proc. Optical Fiber Communication Conference* (OFC'11), Los Angeles (CA), USA, Paper OML4, March 2011

[C5]   S. Koenig, M. Hoh, **R. Bonk**, H. Wang, P. Pahl, T. Zwick, C. Koos, W. Freude, and J. Leuthold, "Rival Signals in SOA Reach-Extended WDM-TDM-GPON Converged with RoF," *Proc. Optical Fiber Communication Conference* (OFC'11), Los Angeles (CA), USA, Paper OWT1, March 2011

[C6]   W. Freude, D. Hillerkuss, T. Schellinger, R. Schmogrow, M. Winter, T. Vallaitis, **R. Bonk**, A. Marculescu, J. Li, M. Dreschmann, J. Meyer, S. Ben Ezra, M. Caspi, B. Nebendahl, F. Parmigiani, P. Petropoulos, B. Resan, A. Oehler, K. Weingarten, T. Ellermeyer, J. Lutz, M. Möller, M. Huebner, J. Becker, C. Koos, and J. Leuthold, "Orthogonal frequency division multiplexing (OFDM) in photonic communications," *10th Intern. Conf. on Fiber Optics & Photonics* (Photonics'08), Indian Institute of Technology Guwahati (IIT Guwahati), Guwahati, Assam, India, December 11-15, 2010 [invited]

[C7]   W. Freude, L. Alloatti, T. Vallaitis, D. Korn, D. Hillerkuss, **R. Bonk**, R. Palmer, J. Li, T. Schellinger, M. Fournier, J. Fedeli, W. Bogaerts, P. Dumon, R. Baets, A. Barklund, R. Dinu, J. Wieland, and M. L. Scimeca, "High-speed signal processing with silicon-organic hybrid devices," *European Optical Society Annual Meeting* (EOS'10), Parc Floral De Paris, France, October 2010 [invited]

[C8]   L. Alloatti, D. Korn, D. Hillerkuss, T. Vallaitis, J. Li, **R. Bonk**, R. Palmer, T. Schellinger, A. Barklund, R. Dinu, J. Wieland, M. Fournier, J. Fedeli, P. Dumon, R. Baets, C. Koos, W. Freude, and J. Leuthold, "40 Gbit/s Silicon-Organic Hybrid (SOH) Phase Modulator," *ECOC 2010*, Torino, Italy, Paper Tu.5.C.4, September 2010

[C9]   S. Koenig, J. Pfeifle, **R. Bonk**, T. Vallaitis, C. Meuer, D. Bimberg, C. Koos, W. Freude, and J. Leuthold, "Optical and Electrical Power Dynamic Range of Semiconductor Optical Amplifiers in Radio-over-Fiber Networks," *ECOC 2010*, Torino, Italy, Paper Th.10.B.6, September 2010

[C10] J. Leuthold, M. Winter, W. Freude, C. Koos, D. Hillerkuss, T. Schellinger, R. Schmogrow, T. Vallaitis, **R. Bonk**, A. Marculescu, J. Li, M. Dreschmann, J. Meyer, M. Huebner, J. Becker, S. Ben Ezra, N. Narkiss, B. Nebendahl, F. Parmigiani, P. Petropoulos, B. Resan, A. Oehler, K. Weingarten, T. Ellermeyer, J. Lutz, and M. Möller, "All-optical FTT signal processing of a 10.8 Tb/s single channel OFDM signal," *Photonics in Switching* (PS), OSA Technical Digest (CD) (Optical Society of America), paper PWC1, July 25-28, 2010 [invited]

[C11] W. Freude, **R. Bonk**, T. Vallaitis, A. Marculescu, A. Kapoor, E. K. Sharma, C. Meuer, D. Bimberg, R. Brenot, F. Lelarge, G.-H. Duan, and . Leuthold, "Linear and nonlinear semiconductor optical amplifiers," *Proc. 12th Intern. Conf. on Transparent Optical Networks* (ICTON'10), München, Germany, Paper We.D4.1, June 27-July 1, 2010 [invited]

[C12] L. Alloatti, D. Korn, D. Hillerkuss, T. Vallaitis, J. Li, **R. Bonk**, R. Palmer, T. Schellinger, A. Barklund, R. Dinu, J. Wieland, M. Fournier, J. Fedeli, W. Bogaerts, P. Dumon, R. Baets, C. Koos, W. Freude, and J. Leuthold, "Silicon High-Speed Electro-Optic Modulator," *Group IV Photonics 2010*, China, paper ThC2 Sept. 2010

[C13] T. Vallaitis, **R. Bonk**, J. Guetlein, C. Meuer, D. Hillerkuss, W. Freude, D. Bimberg, and J. Leuthold, "Optimizing SOA for large input power dynamic range with respect to applications in extended GPON," in *OSA Topical Meeting: Access Networks and In-house Communications* (ANIC), 2010, paper AThC4

[C14] M. Spyropoulou, **R. Bonk**, D. Hillerkuss, N. Pleros, T. Vallaitis, W. Freude, I. Tomkos, and J. Leuthold, "Experimental investigation of multi-wavelength clock recovery based on a quantum-dot SOA at 40 Gb/s," in *OSA Topical Meeting: Signal Processing in Photonic Communications* (SPPCom), Karlsruhe, Germany, Jun. 21-24 2010, paper SPTuB4

[C15] D. Hillerkuss, T. Schellinger, R. Schmogrow, M. Winter, T. Vallaitis, **R. Bonk**, A. Marculescu, J. Li, M. Dreschmann, J. Meyer, S. B. Ezra, N. Narkiss, B. Nebendahl, F. Parmigiani, P. Petropoulos, B. Resan, K. Weingarten, T. Ellermeyer, J. Lutz, M. Möller, M. Huebner, J. Becker, C. Koos, W. Freude, and J. Leuthold, "Single source optical OFDM transmitter and optical FFT receiver demonstrated at line rates of 5.4 and 10.8 Tbit/s," in *Proc. Optical Fiber Communication Conference* (OFC'10), San Diego, CA, USA, Mar. 21–25 2010, postdeadline paper PDPC1

[C16] J. Leuthold, **R. Bonk**, T. Vallaitis, A. Marculescu, W. Freude, C. Meuer, D. Bimberg, R. Brenot, F. Lelarge, and G.-H. Duan, "Linear and nonlinear semiconductor optical amplifiers," in *Proc. Optical Fiber Communication Conference* (OFC'10), San Diego, CA, USA, Mar. 21-25 2010, invited paper

[C17] T. Vallaitis, **R. Bonk**, J. Guetlein, D. Hillerkuss, J. Li, W. Freude, J. Leuthold, C. Koos, M. L. Scimeca, I. Biaggio, F. Diederich, B. Breiten, P. Dumon, and R. Baets, "All-optical wavelength conversion of 56 Gbit/s NRZ-DQPSK signals in silicon-organic hybrid strip waveguides," in *Proc. Optical Fiber Communication Conference* (OFC'10), San Diego, CA, USA, Mar. 21-25 2010, paper OTuN1

[C18] J. Leuthold, **R. Bonk**, P. Vorreau, S. Sygletos, D. Hillerkuss, W. Freude, G. Zarris, D. Simeonidou, C. Kouloumentas, M. Spyropoulou, I. Tomkos, F. Parmigiani, P. Petropoulos, D. J. Richardson, R. Weerasuriya, S. Ibrahim, A. D. Ellis, R. Morais, P. Monteiro, S. Ben Ezra, and S. Tsadka, "All-optical grooming for 100 Gbit/s Ethernet," SPIE Photonics West, San Francisco (CA), USA, Paper 7621-07, Jan. 2010 [invited]

[C19] **R. Bonk**, T. Vallaitis, J. Guetlein, D. Hillerkuss, J. Li, W. Freude, and J. Leuthold, "Quantum dot SOA dynamic range improvement for phase modulated signals," in *Proc. Optical Fiber Communication Conference* (OFC'10), San Diego, CA, USA, Mar. 21-25 2010, paper OThK3

[C20] W. Freude, **R. Bonk**, T. Vallaitis, A. Marculescu, A. Kapoor, C. Meuer, D. Bimberg, R. Brenot, F. Lelarge, G.-h. Duan, and J. Leuthold, "Semiconductor optical amplifiers (SOA) for linear and nonlinear applications," *in DPG Spring Meeting 2010*, Regensburg, Germany, Mar. 2010, Topical Talk DS2.3

[C21] S. Sygletos, **R. Bonk**, T. Vallaitis, A. Marculescu, P. Vorreau, J. Li, R. Brenot, F. Lelarge, G. Duan, W. Freude, and J. Leuthold, "Optimum filtering schemes for performing wavelength conversion with QD-SOA," in *11th International Conference on Transparent Optical Networks, ICTON '09*, Jun. 28-Jul. 2 2009, invited paper Mo.C1.3

[C22] J. Leuthold, **R. Bonk**, P. Vorreau, S. Sygletos, D. Hillerkuss, W. Freude, G. Zarris, D. Simeonidou, C. Kouloumentas, M. Spyropoulou, I. Tomkos, F. Parmigiani, P. Petropoulos, D. J. Richardson, R. Weerasuriya, S. Ibrahim, A. D. Ellis, C. Meuer, D. Bimberg, R. Morais, P. Monteiro, S. Ben Ezra, and S. Tsadka, "An all-optical grooming switch with regenerative capabilities," *Proc. 11th Intern. Conf. on Transparent Optical Networks* (ICTON'09), Ponta Delgada, Island of São Miguel, Portugal, Paper We.A3.4, Vol. 3, pp. 1-4, June 28-July 2, 2009 [invited]

[C23] T. Vallaitis, D. Hillerkuss, J.-S. Li, **R. Bonk**, N. Lindenmann, P. Dumon, R. Baets, M. L. Scimeca, I. Biaggio, F. Diederich, C. Koos, W. Freude, and J. Leuthold, "All-optical wavelength conversion using cross-phase modulation at 42.7 Gbit/s in silicon-organic hybrid (SOH) waveguides," in *Proc. Photonics in Switching*, 2009 (PS 2009). International Conference on, Pisa, Italy, Sep. 15-19 2009, postdeadline paper PD3

[C24] T. Vallaitis, C. Heine, **R. Bonk**, W. Freude, J. Leuthold, C. Koos, B. Esembeson, I. Biaggio, T. Michinobu, F. Diederich, P. Dumon, and R. Baets, "All-optical wavelength conversion at 42.7 Gbit/s in a 4 mm long silicon-organic hybrid waveguide," in *Conference on Optical Fiber Communication 2009*, OFC 2009, San Diego, CA, USA, Mar. 22-26 2009, paper OWS3

[C25] **R. Bonk**, R. Brenot, C. Meuer, T. Vallaitis, A. Tussupov, J. C. Rode, S. Sygletos, P. Vorreau, F. Lelarge, G.-H. Duan, H.-G. Krimmel, T. Pfeiffer, D. Bimberg, W. Freude, and J. Leuthold, "1.3 / 1.5 µm QD-SOAs for WDM/TDM GPON with extended reach and large upstream / downstream dynamic range," in *Conference on Optical Fiber Communication 2009*, OFC 2009, San Diego, CA, USA, Mar. 22-26 2009, paper OWQ1

[C26] G. Zarris, F. Parmigiani, E. Hugues-Salas, R. Weerasuriya, D. Hillerkuss, N. Amaya Gonzalez, M. Spyropoulou, P. Vorreau, R. Morais, S. K. Ibrahim, D. Klonidis, P. Petropoulos, A. D. Ellis, P. Monteiro, A. Tzanakaki, D. Richardson, I. Tomkos, **R. Bonk**, W. Freude, J. Leuthold, and D. Simeonidou, "Field trial of WDM-OTDM transmultiplexing employing photonic switch fabric-based buffer-less bit-interleaved data grooming and all-optical regeneration," *Optical Fiber Communication Conference* (OFC'09), San Diego (CA), USA, 22.-26.03.2009. Postdeadline Paper PDPC10

[C27] C. Schmidt-Langhorst, C. Meuer, R. Ludwig, D. Puris, **R. Bonk**, T. Vallaitis, D. Bimberg, K. Petermann, J. Leuthold, and C. Schubert, "Quantum-dot semiconductor optical booster amplifier with ultrafast gain recovery for pattern-effect free amplification of 80 Gb/s RZ-OOK data signals," in *35th European Conference on Optical Communication, ECOC 2009*, Vienna, Austria, Sep. 20-24 2009, paper We6.2.1.

[C28] **R. Bonk**, S. Sygletos, R. Brenot, T. Vallaitis, A. Marculescu, P. Vorreau, J. Li, W. Freude, F. Lelarge, G.-H. Duan, and J. Leuthold, "Optimum filter for wavelength conversion with QD-SOA," in *Proc. Conf. on Lasers and Electro-Optics* (CLEO/IQEC 2009), Baltimore, USA, May 31-Jun. 05 2009, paper CMC6

[C29] J. Leuthold, W. Freude, S. Sygletos, **R. Bonk**, T. Vallaitis, and A. Marculescu, "All-optical regeneration," in *Asia Communications and Photonics Conference and Exhibition*, Shanghai, China, Nov. 2-6 2009, invited paper TuK1

[C30] N. Lindenmann, T. Vallaitis, **R. Bonk**, C. Koos, W. Freude, and J. Leuthold, "Amplitude and phase dynamics in silicon compatible waveguides with highest Kerr-nonlinearities," in *AMOP Spring Meeting, Deutsche Physikalische Gesellschaft*, DPG Quantum Optics and Photonics Section, Hamburg, Germany, Mar. 2-6 2009, paper Q 57.4

[C31] S. Sygletos, **R. Bonk**, P. Vorreau, T. Vallaitis, J. Wang, W. Freude, J. Leuthold, C. Meuer, D. Bimberg, R. Brenot, F. Lelarge, and G.-H. Duan, "A wavelength conversion scheme based on a quantum-dot semiconductor optical amplifier and a delay interferometer," in *10th Anniversary International Conference on Transparent Optical Networks*, ICTON 2008, Athens, Greece, Jun. 22-26 2008, invited paper We.B2.5

[C32] J. Leuthold, W. Freude, S. Sygletos, P. Vorreau, **R. Bonk**, D. Hillerkuss, I. Tomkos, A. Tzanakaki, C. Kouloumentas, D. J. Richardson, P. Petropouos, F. Parmigiani, A. Ellis, D. Cotter, S. Ibrahim, and R. Weerasuriya, "An All-Optical Grooming Switch to Interconnect Access and Metro Ring Networks," *Proc. 10th International Conference on Transparent Optical Networks* (ICTON'2008), Athens, Greece, paper We.C3.4, June 2008

[C33] **R. Bonk**, P. Vorreau, S. Sygletos, T. Vallaitis, J. Wang, W. Freude, J. Leuthold, R. Brenot, F. Lelarge, G.-H. Duan, C. Meuer, S. Liebich, M. Laemmlin, and D. Bimberg, "An interferometric configuration for performing cross-gain modulation with improved signal quality," in *Conference on Optical Fiber communication/National Fiber Optic Engineers Conference*, OFC/NFOEC 2008, San Diego, CA, USA, Feb. 24-28 2008, paper JWA70

[C34] **R. Bonk**, C. Meuer, T. Vallaitis, S. Sygletos, P. Vorreau, S. Ben-Ezra, S. Tsadka, A. Kovsh, I. Krestnikov, M. Laemmlin, D. Bimberg, W. Freude, and J. Leuthold, "Single and multiple channel operation dynamics of linear quantum-dot semiconductor optical amplifier," in *34th European Conference on Optical Communication*, ECOC 2008, Brussels, Belgium, Sep. 21-25 2008, paper Th.1.C.2

[C35] **R. Bonk**, S. Sygletos, P. Vorreau, T. Vallaitis, J. Wang, W. Freude, J. Leuthold, R. Brenot, F. Lelarge, and G.-H. Duan, "Performance analysis of an interferometric scheme for media with limited cross-phase modulation nonlinearity," in *Proc. 6th Intern. Symp. on Communication Systems, Networks and Digital Signal Processing* (CNSDSP'08), Graz, Austria, Jul. 25-25 2008, pp. 487-491

[C36] P.Vorreau, D. Hillerkuss, F. Parmigiani, S. Sygletos, **R. Bonk**, P. Petropoulos, D. Richardson, G. Zarris, D. Simeonidou, D. Klonidis, I. Tomkos, R. Weerasuriya, S. Ibrahim, A. Ellis, R. Morais, P. Monteiro, S. Ben Ezra, S. Tsadka, W. Freude, and J. Leuthold, "2R/3R Optical Grooming Switch with Time-Slot Interchange," *Proc. of 34rd European Conference on Optical Communication*, (ECOC'08), Brüssel, Sept. 2008

[C37] C. Meuer, M. Laemmlin, S. Liebich, J. Kim, D. Bimberg, A. Capua, G. Eisenstein, **R. Bonk**, T. Vallaitis, and J. Leuthold, "High speed cross gain modulation using quantum dot semiconductor optical amplifiers at 1.3μm," in *Proc. Conf. on Lasers and Electro-Optics* (CLEO/IQEC 2008), San Jose, CA, USA, May 4-9 2008, paper CTuH2

[C38] J. Leuthold, J. Wang, T. Vallaitis, C. Koos, **R. Bonk**, A. Marculescu, P. Vorreau, S. Sygletos, and W. Freude, "New approaches to perform all-optical signal regeneration," in *9th International Conference on Transparent Optical Networks*, ICTON '07, Rome, Italy, Jul. 1-5 2007, invited paper We.D2.1

[C39] S. Sygletos, M. Spyropoulou, P. Vorreau, **R. Bonk**, I. Tomkos, W. Freude, and J. Leuthold, "Multi-wavelength regenerative amplification based on quantum-dot semiconductor optical amplifiers," *Proc. 9th Intern. Conf. on Transparent Optical Networks* (ICTON'07), Rome, Italy, Paper We.D2.5, pp. 234-237 (invited) July 2007

[C40] T. Vallaitis, C. Koos, B.-A. Bolles, **R. Bonk**, W. Freude, M. Laemmlin, C. Meuer, D. Bimberg, and J. Leuthold, "Quantum dot semiconductor optical amplifier at 1.3μm for ultra-fast cross-gain modulation," in *33th European Conference on Optical Communication*, ECOC 2007, Berlin, Germany, Sep. 16-20 2007, paper We8.6.5

[C41] C. Koos, T. Vallaitis, B.-A. Bolles, **R. Bonk**, W. Freude, M. Laemmlin, C. Meuer, D. Bimberg, A. D. Ellis, and J. Leuthold, "Gain and phase dynamics in an InAs/GaAs quantum dot amplifier at 1300 nm," in *Proc. Conf. on Lasers and Electro-Optics* (CLEO/IQEC 2007), Munich, Jun. 17-22 2007, paper CI3-1-TUE

[C42] J. Wang, Y. Jiao, **R. Bonk**, W. Freude, and J. Leuthold, "Regenerative Properties of Bulk and Quantum Dot SOA Based All-Optical Mach-Zehnder Interferometer DPSK Wavelength Converters," *Proc. International Conference on Photonics in Switching 2006*, Herakleion (Crete), Greece, Oct. 2006

[C43] J. Leuthold, W. Freude, G. Boettger, J. Wang, A. Marculescu, P. Vorreau, and **R. Bonk**, "All-Optical Regeneration," *Proc. 8th Intern. Conf. on Transparent Optical Networks* (ICTON'06), Nottingham, UK, June 2006 (invited)

Curriculum Vitae

René Bonk
renebonk@msn.com

Work Experience

04/2006–03/2012   **Karlsruhe   Institute   of   Technology   (KIT), Institute of Photonics and Quantum Electronics (IPQ)**
PhD studies
- Communication system testing, optimization and development of semiconductor waveguides devices (e.g., quantum-dot semiconductor optical amplifier (QD SOA)) for linear applications at highest speeds, advanced modulation formats and multiplexing techniques
- Experimental demonstration of ultra-fast nonlinear all-optical signal processing in a communication system context based on QD SOA
- Coordinator of the European project TRIUMPH (Transparent Ring Interconnection Using Multi-Wavelength Photonic Switches), which included the implementation of an all-optical high-speed grooming switch
- Contribution to work packages and deliverables of several European and national research projects
- Tutorial lectures on optical communications (fiber optics and waveguides, lasers, detectors, communication systems, nonlinear optics)
- Supervision of student projects and master theses

03/2003–07/2005   **Technical University of Braunschweig (TH)**
- Advisor of lab studies for electro-engineering students

Education

04/2006–present	**University of Karlsruhe (TH) / Karlsruhe Institute of Technology (KIT), Institute of Photonics and Quantum Electronics (IPQ)**
	PhD thesis on "Linear and Nonlinear Semiconductor Optical Amplifiers for Next-Generation Optical Networks"
	Date of defense: April 19, 2012.
01/2006	**Technical University of Braunschweig**
	Major in Physics, "Dipl.-Phys."
	Grade 1.0 (sehr gut, "very good")
01/2005–01/2006	**Technical University of Braunschweig**
	Applied Physics Department
	Final thesis "Terahertz photodetectors based on HgTe/HgCdTe heterostructures"
	Grade: 1.0 (sehr gut, "very good")
10/2000–12/2004	**Technical University of Braunschweig**
	Major in Physics:
	- Semiconductor physics
09/1999–07/2000	**Military service**
08/1992–07/1999	**Gymnasium Uelzen**
	(secondary school)
	Allgemeine Hochschulreife (general qualification for university entrance)
08/1986–07/1992	**Grundschule and Orientierungsstufe Bad Bodenteich**
	(elementary school)

Karlsruhe Series in Photonics & Communications
KIT, Institute of Photonics and Quantum Electronics (IPQ)
(ISSN 1865-1100)

Die Bände sind unter www.ksp.kit.edu als PDF frei verfügbar
oder als Druckausgabe bestellbar.

12801492R00167

Made in the USA
Monee, IL
29 September 2019